Unifying Biology

Unifying Biology

THE EVOLUTIONARY SYNTHESIS AND
EVOLUTIONARY BIOLOGY

Vassiliki Betty Smocovitis

PRINCETON UNIVERSITY PRESS

PRINCETON, NEW JERSEY

Copyright © 1996 by Princeton University Press
Published by Princeton University Press, 41 William Street,
Princeton, New Jersey 08540
In the United Kingdom: Princeton University Press,
Chichester, West Sussex

Library of Congress Cataloging-in-Publication Data

Smocovitis, Vassiliki Betty.
Unifying biology : the evolutionary synthesis and
evolutionary biology / Vassiliki Betty Smocovitis.
p.   cm.
Includes bibliographical references and index.
ISBN 0-691-03343-9 (cloth : alk. paper)
1. Evolution (Biology)—History.
2. Biology—History. I. Title.
QH361.S64   1996
575'.009—dc20   96-5605

This book has been composed in Galliard

Princeton University Press books are printed on
acid-free paper and meet the guidelines for permanence
and durability of the Committee on Production
Guidelines for Book Longevity of the
Council on Library Resources

Printed in the United States of America
by Princeton Academic Press

10 9 8 7 6 5 4 3 2 1

To Ernst Mayr and William B. Provine

His father studied him for a long time in bewilderment before saying, "For this you want to sacrifice your future? The origin of life and its destiny? The origin of life was Adam, and our destiny is paradise and hellfire. Or has there been some new discovery concerning this?"

Naguib Mahfouz, *The Palace of Desire, The Cairo Trilogy II*

The doors of heaven and hell are adjacent and identical.

Nikos Kazantzakis, *The Last Temptation of Christ*

Contents

Preface

> A congenital synthesizer, I held on to the dream of a unify-
> ing theory.
>
> Edward O. Wilson, *Naturalist*

THIS book is concerned with history, coherence, and narratives, especially the power of narratives to lend coherence to intellectual and scientific projects. It is especially concerned with grand narratives—all-embracing, universalizing, originary stories told about the universe, life and
humanity, and about the construction of the grandest narrative of Western culture, the modern story of evolution.

The project is also about the struggle to lend coherence and to unify
the most heterogeneous of all the sciences, biology, within an Enlightenment worldview. It historically recovers this struggle through the emergence of the unifying discipline of evolutionary biology. At the same
time, it begins to explore one of the most central and enigmatic episodes
in the history of biology, the evolutionary synthesis. The project is also
about the desire that drives such unificatory projects—in the face of the
adversity of diversity—and about the modern culture of science, a historically rooted expression that would not exist without passions, dreams,
and desires.

It is also very much a personal project grown out of the chaotic world of
lived experience with the diversity of the biological sciences in our modern world. Written with all the advantages that hindsight affords us, the
following background narrative to the project makes for a much more
coherent account than would otherwise be possible. Such logic also ap-
plies to prefaces, which come first in a book, but are clearly written last.

THE STORY OF THE PROJECT

This book grew out of a seminar in the history of biology taught in
the spring of 1982 at Cornell University by William B. Provine. Intended for advanced undergraduate and graduate students in biology,
history, and philosophy, the seminar was designed to explore the "persistent controversies" that ensued following the publication of Darwin's *On
the Origin of Species* in 1859. Although the course covered late-nine-

teenth-century developments leading to the "eclipse of Darwin" (Julian Huxley's well-worn phrase) around the year 1900, it focused on developments in evolution in the years between 1920 and 1950. It was at this time that the critical event Provine called the "evolutionary synthesis" took place. According to existing historical understanding, the synthesis involved the integration of Darwinian selection theory with the newer science of genetics, leading to the establishment of the Neo-Darwinian or modern synthetic theory of evolution. In addition to reviving Darwinism and modernizing evolution, the synthesis was also thought to have brought rival factions, primarily naturalist-systematists and experimental geneticists, together through new developments in mathematical or theoretical population genetics. One result was an evolutionary framework that integrated systematics, paleontology, botany, and other sciences within what appeared to be a unified evolutionary theory. The bulk of the course was devoted to study of the texts, fields, relevant problems, and contributions of individuals toward understanding the historical conditions that led to the synthesis.

But almost immediately after such introductory and rather general characterizations of the episode could be made, historical understanding of the synthesis was called into question. What exactly was meant by synthesis? What individuals and what fields had contributed the most to the synthesis? Were any new facts about evolution gained, or were the old ones merely reshuffled? What was the proper meaning of the synthesis? In short time, it was revealed to the class that so many such questions had been raised and had failed to be answered, that scholars devoting their energy to the subject had failed to reach any consensus. So many qualifiers, questions, and perspectives were introduced, that the students quickly echoed historians who recognized it as one of the most vexing problems in the history of biology.

Despite its formidable demands in the way of technical knowledge of evolution (and a demanding set of readings) the course drew a sizable number of students who had already gained appreciation of the historical, philosophical, or social study of biological science. That the course had drawn such a number of able students, and that it had been taught at Cornell University under the direction of Provine at that time was understandable. A unique administrative hybrid between a state school and a private, endowed institution, Cornell had combined some of the best of both worlds, supporting the diverse areas of biology that were proliferating in the United States. Unlike many other major research universities, which, either through deliberate or inadvertent decision making, had chosen to support some areas of biological research to the exclusion of others, Cornell's "double vision" permitted the inclusion of a greater number of biological sciences than was usual. In addition to keeping up with the newer sciences of cellular, biochemical, and molecular biology, Cornell preserved its tradition in natural history and systematics, as well as investing heavily in life sciences that connected to agriculture, natural resources, education, and veterinary medicine (only the medical school was located off-campus in New York City). Thus, because of its history, research emphasis, and administrative setup, Cornell had become a leading institution representing the diversity and heterogeneity of the modern biological sciences.

Both undergraduate and graduate study in biological sciences offered not only a diverse curriculum but also flexible and innovative ways to override the narrowness of intradepartmental or monodisciplinary study. This was especially true at the graduate level, which placed students in "fields" that did not necessarily coincide with departments, "sections," or the location of their graduate committee. As a result, graduate students at Cornell were especially free to design their program of study and could easily cross traditional departmental boundaries. The seminar that Provine offered crossed into so many areas and fields that it drew a sizable, yet diverse, assemblage of interested students beyond what one would expect for an advanced seminar.

Provine's course was also timed to take advantage of the new volume on the evolutionary synthesis that he had completed editing with the noted evolutionary biologist, historian, and philosopher at Harvard University, Ernst Mayr. Mayr had also been one of the original participants or "architects" (as they came to be known) of the evolutionary synthesis. The appearance of the completed volume on this important subject in 1980 begged for seminar-like group discussion, especially given the numerous advanced undergraduate and graduate students at Cornell interested in the history of evolution and biology. Although it hardly resembled a conventional textbook of instruction, Ernst Mayr and William B. Provine's edited volume, *The Evolutionary Synthesis: Perspectives on the Unification of Biology* served as primary guide and resource for the seminar.

Provine proved to be a charismatic and energetic teacher, drawing scores of students to this and his other classes. His own location at the time was in the department of history, though he held adjunct appointments in the section of ecology and systematics. Because his research melded modern evolutionary theory with historical insights, and because his work spoke to scientists, Provine crossed the great divide between the humanities and sciences by transferring his location full-time to the section of ecology and systematics, while at the same time keeping his appointment in history. From the mid-1980s on Provine was able to cross disciplinary and departmental boundaries by training graduate students in both the department of history and the section of ecology and systematics and eventually in the Program in the History and Philosophy of Science and Technology.

THAT spring of 1982 I was a new graduate student at Cornell in the field of "Ecology and Evolutionary Biology." Provine's course, which examined the historical background to evolution, seemed relevant to my interests, especially as I then believed that all really good scientists had to know something about the history of their field. Though my interest was in evolutionary biology, I had entered Cornell through the administrative entity of the section of plant biology. This was in part because of a previous background in botany and the plant sciences, but also because it seemed that all biologists, almost by definition, had to have *some* organism(s) to study. I had also chosen to study paleobotany, the science that most immediately explored the historical or evolutionary record of plant life (a fascination with reconstructions of giant horsetails and lycopods was another reason for choice of study). I had thus entered graduate school with the goal of understanding evolution and with the hope of becoming an evolutionary biologist.

Actual experience of the diversity of the biological sciences did not make for the most coherent course of study, however, as I quickly discovered. The curious organization of biology into sections (rather than departments), the even stranger organization of "graduate fields" (within which one chose a major area), the variable location of the graduate advisor, and the overarching split between the statutory and endowed portions of the university had the potential of creating administrative horrors for those individuals who did not—and could not—have a clear alignment among advisor, department, section, and graduate field. Equally difficult to negotiate were the differences of views among the biologists on the most fundamental of scientific points, brought out repeatedly by the unusual groupings and regroupings that the flexibility of the system permitted. This became especially difficult as the number of faculty increased over the years and spread out into numerous buildings. The "Division of Biological Sciences" (this seemed to me then an especially suitable rubric) was actually so complicated that a special large poster in the form of an organizational flowchart with photographs of faculty was printed, distributed, and then displayed widely along corridors, bulletin boards, and offices to assist members in getting to know the faces, fields, and sections and how they fit into the "Division of Biological Sciences."

In addition to the intellectual dissonance brought out by the organization of the "Division," and the constant danger of falling between the administrative cracks in important matters concerning teaching assignments and financial support, graduate students in "Ecology and Evolutionary Biology" (E&EB) had other problems. E&EB was not only one of the larger graduate fields of biology, making it difficult to manage, but also by virtue of its very subject matter appeared amorphous and diffusely organized. Representative faculty could be found in at least eight different buildings and in at least five different sections/departments, between statutory and endowed portions of the university. Relevant courses in evolution were offered not only in the biological sections, but also in departments of geology, anthropology, chemistry, and the space sciences (along with historical, philosophical, and science and society courses). Two years of required graduate study within this plan left students potentially traversing the campus across all manner of sections, departments, and buildings in search of relevant courses and faculty as they studied aspects of evolution, whether cosmic, chemical, molecular, human, or cultural, along with the more familiar "organic" evolution taught in biology.

Students daring the crossings could not help but notice major differences of opinion in the content of the courses. These differences ranged from the aims and purposes of study, to the most appropriate methods and instruments of study, to the proper unit or level of evolution, to the predominant mechanism operant in evolution, to the most appropriate organism of study. Conflicts and debates introduced in one course were ignored in another as the most basic of definitional parameters like the meaning of terms such as "fitness," "adaptation," and even "evolution" came into question. No course seemed to do justice to the entirety of the evolutionary process and no one teacher could effectively address the differences seen by students as they made their way through the individual subject matter. A graduate "core course in evolution" drawing all students together with relevant faculty seemed an impossibility.

The wedding of "Ecology with Evolutionary Biology" (a very popular combination at American universities) also made for an occasionally volatile mix, especially when graduate field categories such as E&EB tried to fit within the section (or the administrative unit) of ecology and systematics. Not all evolutionary biologists were systematists and vice versa. In one heated discussion proponents of systematics argued that it actually subsumed evolutionary biology; this had been deemed sufficient reason *not* to change the section name to correspond with the graduate field of "Ecology and Evolutionary Biology." For similar reasons, there were also occasional moments of tension between those individuals more in "E" and those leaning toward "EB." Students of the latter did not necessarily share the same attitudes, values, aims, and habits of the former, nor did they always take a similar course of study; few ecologists traversed the campus to explore the range of courses and faculty in evolution, for instance, and some of us in "EB" had little or no experience with field ecology. The differences between the two were foregrounded by the noninteractions between ecosystem ecologists and more organismally inclined evolutionists (some of the former cared little for individual organisms in nature and some of the latter cared little for mathematical models dangling in hyperspace). One especially heated exchange in a graduate seminar in E&EB polarized discussants between the subsuming rhetoric of ecologists and the few of us in evolution left chanting "Nothing in biology makes sense except in the light of evolution" (Theodosius Dobzhansky's oft-heard assertion in our "intro bio" texts). In so doing we counterargued with an equally, if not more subsuming rhetorical position that had been handed down to us by the very same architects of the evolutionary synthesis who claimed that evolution somehow "unified" the whole of biology (a rather ambitious act of faith given our experiences with the diversity of the biological sciences).

As if the administrative couplings of ecology, systematics, evolution, and biology and the very setup at Cornell were not problematic enough, the current scientific debates centering on evolution only added to the confusion surrounding its study. The early 1980s—as evolutionists who witnessed these years know—was an exceedingly tumultuous time for study of evolution. At the same time that these years generated excitement and drew interest to evolutionary biology, they also served to divide evolutionists on methodological, institutional, political, and personal grounds. The most fundamental of points in evolution came into question as evolutionists challenged the very synthetic theory that Provine's course had explored in such detail; the scientific edifice of the evolutionary synthesis itself appeared to crack under the strain.

Two and one half years into this program of study I was even more convinced that the more I learned about evolution, the less I knew (a sure sign of pending Ph.D. candidacy, I was told). As the educational process of negotiating these conflicts and sorting through the content and organization of evolution, ecology, systematics, and the remainder of biology against the backdrop of the physical sciences played itself out, my interests shifted from actually doing evolution to focusing on evolution as a subject worthy of study in itself. Delving into the history and philosophy of evolutionary biology, it also became increasingly obvious that the "problem" of the synthesis was intimately wrapped up with the diffuse organization of evolution and biology. Questions thus shifted: What exactly was evolutionary biology? What were its historical origins? What role did the evolutionary synthesis play in evolutionary biology? And why exactly was the synthesis of such historic importance? Could biology really claim to be a unified science, given its obvious heterogeneity? And why was there a widespread belief that evolution served to unify the whole of biology, especially in the face of more molecular practitioners? From firsthand experience, evolution seemed to consist of a heterogeneous complex of subdisciplines; so, too, was this the case for the wider biological sciences. Yet despite the heterogeneity, belief in a sense of unity among biologists—and certainly evolutionists—existed (even though biologists had demonstrated the tendency to overstress the importance and location of their own immediate area of expertise). Nearly all biologists believed themselves to be part of the same project, and all ultimately sought a coherent unifying theory of life commensurate with the laws of physics and chemistry. For most practitioners, if the biological sciences appeared incoherent, this was an artifact of organization and administration, rather than of the science. Nearly all ultimately believed they were engaged in a unificatory project that followed one scientific method and that led to the most truthful claims about the natural world (if not the absolute truth).

An experience with teaching "writing-across-disciplines" in the John S. Knight Writing Program in the English department at Cornell challenged all such beliefs as I encountered countervailing attitudes prevalent in the humanities. According to their theoretical perspective drawn from cultural anthropologists like Clifford Geertz, disciplines of knowledge were cultures of their own, complete with rituals, attitudes, and beliefs. Entry or enculturation into the discipline involved knowledge and use of what was termed "disciplinary discourse"; hence the goal of teaching students to write, or "control" the language of their discipline, rather than having the language control them, was a critical part of their education within the discipline. In the process students also grew to experience, appreciate, and respect the "multiculturalism" inherent in the diverse communities of knowledge.

Though applications of such cultural theory to scientific disciplines had to be qualified, I could also see that the biological sciences at Cornell supported (on the surface at least) such a claim. I also could not help but recognize that scientific disciplines, departments, sections, and other such groupings appeared very much like the cultures that I had experienced in my more personal ethnic peregrinations. At this surface level, therefore, disciplinary/departmental/section crossings were really not unlike those familiar to students in anthropology encountering the problem(s) of culture(s). Among these were included the "other mind" problem, the insider–outsider dilemma, and questions into the social dynamics of cultural cohesion/discontinuity. Could some forms of contemporary cultural theory, sociology of knowledge, or other areas of inquiry within the field defining itself as "science studies" that viewed science as a sociocultural enterprise shed light on evolution and biology? Or was scientific knowledge such a different area of inquiry that cultural theory could not apply? How was the sense of cultural unity attained within cultural diversity? And could the answer to this question help us understand how the sense of scientific unity was attained within so much diversity?

Unifying Biology: The Evolutionary Synthesis and Evolutionary Biology thus took shape against the backdrop of all these experiences that converged in the 1980s. It took well over ten years to weave together a coherent story that could (1) function as an interpretive framework for the evolutionary synthesis, (2) reveal something about the history and organization of the discipline of evolutionary biology, and (3) understand how these played into the growing belief that evolution unified the modern biological sciences. In the process, I had crossed as wide a range of the biological sciences as I could possibly manage, even going so far as to studying exobiology, the one biological science with only a conceivable—but highly probable—object of study.

I had also unexpectedly entered totally unfamiliar terrain as untangling the problem of the synthesis took me through disciplines as diverse as history, philosophy, sociology, anthropology, literary studies, and science studies, ending (I hope) with cultural studies. From each of these disciplines, I have selected some building block or tool to fulfill an "architectural plan" I have designed for the project. *Unifying Biology* should thus be understood within the following series of historiographical or methodological considerations that serve as theoretical scaffolding or as materials and methods for the work: (1) the attempt to write a "cultural" history of science, which views science as contextual, discursive activity; (2) the attempt to develop a view of science that refuses to segregate traditional "internal" from "external" factors; (3) the attempt to respond to—and respect—"postcolonial ethnographic" demands that acknowledge or give voice to the perspective of the historical actors; (4) the attempt to situate my own authorial voice or location within the text; (5) the attempt to step "outside" the culture of science, by using tools and methods to create critical distance; (6) the attempt to explore the narrativity of historical sciences like evolution; and, finally, (7) the attempt to develop a theory of science and knowledge as expressive of aesthetic and emotive experience, what we may term an "aesthetic epistemology." What is attempted, in short, is a project to bring together relevant knowledge from many areas to bear on the historical subject, within a unified field of view—a sort of synthesis of "the synthesis." The final project is thus not so much cross-disciplinary or interdisciplinary, but *transdisciplinary*, drawing on questions, methods, and ways of thinking from many diverse cultures of knowledge. I hope that I have traveled far enough to reach a critical distance before returning a synoptic, syncretic, and transdisciplinary gaze on evolution, biology, and the historical "problem" of the synthesis.

Acknowledgments

THIS PROJECT clearly owes a great deal to the students, staff, faculty, and administration of the biological sciences at Cornell. Karl J. Niklas, Dominick Paolillo, and L. Pearce Williams bridged the chasm between the sciences and humanities and were positive, directive influences in plant biology and in the history of science. I wish that Gail Rubin could see the project reach its present textual form. Ruth Hamilton served as a wonderfully patient friend and reminder of the "E" in E&EB. Despite initial resistance, I grew to appreciate cultural and literary theory and how it could shed light on scientific disciplines from the introduction to writing-across-disciplines in the John S. Knight Writing Program.

If I could derive so much from my Cornell experience it was because I had been prepared for it by an undergraduate program at the University of Western Ontario that immersed students in the diversity of the biological sciences (special thanks to Dianne Fahselt and the department of plant sciences). A richly diverse, secular, and public Canadian high school founded in the heady optimism of the centennial year (1967) that dared to teach evolution through the Biological Sciences and Curriculum Study (BSCS) also helped shape this study; sadly it has succumbed to social pressures in Canada and has gone the way of a private, French-speaking-only Catholic high school.

I also owe thanks to numerous institutions, audiences, and individuals who have contributed to the final project. Preliminary versions were read at the Committee on the Conceptual Foundations of Science at the University of Chicago, the Center for the Study of Science in Society at Virginia Polytechnic Institute, the Program in the History and Philosophy of Science at the University of California–Davis, the Center for Cultural Study at the University of California–Los Angeles, and the department of philosophy at Wesleyan University; I wish to thank all for their comments. Especially helpful were the participants in "The Narrative Patterns of Scientific Disciplines" workshop at Hebrew University and Tel-Aviv University in the spring of 1992. Don and Molly Verene at the Institute for Vico Studies at Emory University, whom I met under unforgettable circumstances in Tel Aviv, approved a proposal to explore intellectual history and historiography at their NEH Summer Institute on "Vico and Humanistic Knowledge" in the summer of 1993. In addition to providing the guiding image of William Blake's *Fall of Man*, Don and Molly allowed me to recover the ancient meaning of philosophy as the "love of wisdom." Costas Krimbas, Kostas Gavroglou, Aristides Baltas, and their students in the newly founded history and philosophy of science program in Athens, Greece, served as just such philosophical exemplars. Summer workshops and seminars on the history of biology sponsored by the Dibner Institute for the History of Science and held at the Marine Biological Laboratory at Woods Hole made possible intellectual exchanges with colleagues in the history, philosophy, and social study of biology all over the world. Jane Maienschein, who helped make these possible, also generously read a draft of *Unifying Biology* closely and made numerous helpful suggestions. Audiences at meetings of the History of Science Society, American Studies Association, and the International History, Philosophy and Social Studies of Biology (ISHPSSB) also offered feedback.

Numerous archivists and librarians across the country facilitated research: Nancy Boothe with the Julian Huxley Papers at Woodson Research Center at Rice University and Suzy Taraba at the department of special collections at the Joseph Regenstein Library at the University of Chicago provided invaluable assistance. A grant from the NEH Travel to Collections Program (now extinct) made possible an important research trip to the American Philosophical Society Library in Philadelphia. Beth Carroll-Horrocks was an especially knowledgeable guide through all the APS evolution and genetics collections. Libraries at Cornell University, the University of California–Berkeley, the University of California–Davis, Stanford University, and the Marine Biological Laboratory at Woods Hole were all consulted at some time. Libraries at the University of Florida (including George E. Smathers, Education, and Fine Arts and Architecture) were especially helpful in locating materials. Vernon Kisling at Marsden Science Library at the University of Florida deserves special acknowledgment for his assistance.

Much of the reading in cultural history/studies and a significant part of the research and writing for *Unifying Biology* were undertaken while I was Mellon Fellow in the Humanities at Stanford University. I wish to thank members of the program in the history of science: Max Dresden, Peter Galison, Wilbur Knorr, Pierre Noyes, Tim Lenoir, David Stump, Barbara Kataoka, Shannon Temple, the numerous students (especially Hasok Chang, Wendy Lynch, and Sonja Adams), other wandering postdoctoral fellows, and visitors to the program. Thanks also to Keith Michael Baker and Steve Haber in the department of history; and Marcus

Feldman, Ward Watt, Carole Boggs, John Thomas, the numerous students, and other members of "evo"-lunch in the department of biology. Pat Kelly was always "just around the corner" when I needed a friend.

As well as granting generous leave time, the department of history at the University of Florida provided the perfect intellectual environment for transdisciplinary study. An early version of *Unifying Biology* was discussed by the members of Fred Gregory's "breakfast club" at our friendly neighborhood Burger King: Bob D'Amico, Toby Appel, Don Dewsbury, Brandy Kershner, Charlotte Porter, Eldon Turner, and others were critical yet fun readers. Steve Feierman, David Chalmers, Ron Formisano, Pat Geary, Steve Grossbart, Bob Hatch, Susan Kent, Ed McCord, Bob McMahon, Steve McKnight, Louise Newman, Darrett Rutman, Daniel Schroeter, John Sommerville, Jay Tribby, Bob Zeiger, and Bert Wyatt-Brown facilitated work in some way. Harry Paul provided food for nourishment and thought; wine for amusement and anesthesia. Members of the department of English sharing the fourth floor of Turlington Hall (especially David Locke, Andy Gordon, and Dan Cottom) shared reading lists and offered critical exchanges in the hallways. Members of the departments of botany and zoology served as helpful audiences and were especially supportive of the history of biology at UF. Brian Keith McNab demonstrated his admirable hidden skills as an especially literate variant of *Cavia porcellus* during the final critical phase of writing. Graduate students in the department of history and in HIS 6489 (Cultural History and Study of Scientific Knowledge) read and discussed versions of this manuscript: Jeff Brautigam, Frederick Davis, Neal Doran, Michael Futch, Christopher Koehler, Gary Kroll, and Gary Weisel deserve special thanks for their stimulating comments, encouragement, and considerable insights into the project. Mark Lesney was especially encouraging and made numerous fun and creative suggestions. Kimberley Yocum, Greg Kisling, Paige Porter-Brown, and all the staff in the department of history gave generously of their time, humor, and expertise. Chapter 4 grew out of a workshop entitled "The New Contextualism: Science as Discourse and Culture" held at the University of Florida and funded by the department of history and the Florida Humanities Council. Brian Baigrie, Alison Wylie, Joe Rouse, Andy Pickering, David Stump, and Zori Barkan came together for an intense weekend of wonderfully exciting discussions on science as discourse and culture. An early version of the narrative of *Unifying Biology* appeared in the spring 1992 issue of *Journal of the History of Biology* (vol. 25, pp. 1–65). I am grateful for the editorial support from Everett Mendelsohn,

Shirley Roe, Julia McVaugh, and their staff. Kluwer Academic Publishers gave permission to reproduce most of this work. I am also indebted to those cohorts-turned-colleagues who generously read drafts of *Unifying Biology* and made helpful suggestions for improvement: Joe Ackerman, Mary Bartley, and especially Michael Dietrich. Joe Rouse deserves special acknowledgment for his critical reading and encouragement over the years. Friends just a phone call or E-mail message away have all assisted in some way: Mario Biagioli, Ham Cravens, Michael Dennis, Moti Feingold, Robert Marc Friedman, Steve Fuller, Horace Judson, Laura Lovett, Pat Munday, Alex Pang, Thomas Söderquist, Frank Sulloway, Nadine Weidman, and especially Susan Lindee. Andy Wilson provided conversation and information on intellectual history and historiography. M. J. S. Hodge read and commented on a very early version of the narrative of *Unifying Biology*. Marjorie Grene also read a very early draft and provided invaluable long-distance criticism on philosophical matters (though we agree—as always—to disagree). Interviews with G. Ledyard Stebbins toward my dissertation research were instrumental to filling in the wider picture. Pnina Abir-Am, Mark Adams, Gar Allen, John Beatty, Jane Brockmann, Lincoln Brower, Tom Buford, Joe Cain, James Crow, Paul Ewald, Doug Futuyma, Eli Gerson, Stephen Jay Gould, Jon Harwood, Gerald Holton, John Jungck, Paolo Pallodino, Jon Reiskind, Carl Sagan, Marc Swetlitz, Mary Jane West-Eberhardt, Donald Waller, Edward O. Wilson, Leigh Van Valen, and especially John Greene each provided valuable information or facilitated this project in some way. Frank Drake at the University of California–Santa Cruz and the SETI Institute generously shared his life experiences and knowledge of evolutionary cosmology among the Pacific redwoods.

The final phase of writing was completed during a return to Cornell as visiting fellow in ecology and systematics made possible by Will Provine and the Section of Ecology and Systematics. A research/travel grant from the division of sponsored research at UF and a summer research grant from the National Science Foundation made the trip possible. Bill Woodcock at Princeton University Press has been fabulously empathetic to the project and has facilitated its progress to its present form. My parents Alexandra and Dimitrios have made all these influences possible with their unflagging love, support, and encouragement.

My debt to both Will Provine and Ernst Mayr goes without saying. The completion of the narrative ends with the beginning they made possible for a generation of students in *The Evolutionary Synthesis: Perspectives on the Unification of Biology*.

Unifying Biology

CHAPTER ONE

The Exegesis of *Unifying Biology*

> Things taken together are whole and not whole, something
> which is being brought together and brought apart, which is
> in tune and out of tune; out of all things there comes a
> unity, and out of a unity all things.

Heraclitus

> I believe one can divide men into two principal categories:
> those who suffer the tormenting desire for unity, and those
> who do not.

George Sarton

MORE THAN any preface, prologue, or introduction, the image of William Blake's *Fall of Man* immediately captures (powerfully so) the undergirding themes of *Unifying Biology*.[1] On the surface, Blake's image retells the Biblical origin story of Western "Man." Because they partook of the fruit from the tree of knowledge—on Woman's urging—Man and Woman are expelled from the Garden of Eden and sentenced to live between the contrasting worlds of heaven and hell.

Yet the image also carries meanings deeper than this surficial—and somewhat conventional—account of "Man's" origin.[2] In composing the image, Blake studiously divided key components into paired dualities or oppositional images, forming the extremities of the text:[3] God and Satan,

---

[1] William Blake, *Fall of Man*, watercolor, 1807. Printed and discussed as plate 45 in Roger Cook, *The Tree of Life: Image for the Cosmos* (New York: Thames and Hudson, 1974).

[2] For a traditional comparative exegesis of the "Fall of Man," see James G. Frazer, "Fall of Man," in Alan Dundes, ed., *Sacred Narrative* (Berkeley: University of California Press, 1984), pp. 72–97. Reprinted from Sir James G. Frazer, *Folklore in the Old Testament*, vol. 1 (London, 1918), pp. 45–77.

[3] See the discussion of dualism in Blake in Margaret Rudd, *Divided Image: A Study of Blake and W. B. Yeats* (London: Routledge and Kegan Paul, 1953). See also one of the classics in "Blake Studies": Northrop Frye, *Fearful Symmetry: A Study of William Blake* (Princeton: Princeton University Press, 1947). For a discussion of Blake's view of the fall and creation, see J. G. Davies, *The Theology of William Blake* (Oxford: Clarendon, 1948); and Thomas R. Frosch, *The Awakening of Albion: The Renovation of the Body in the Poetry of William Blake* (Ithaca: Cornell University Press, 1974).

Man and Woman, the Tree of Life and the Tree of Knowledge, Good and Evil, Life and Death. These dichotomous elements are obvious to the reader and hardly need closer interpretation; not so obvious, however, is the duality represented by elements in the center and at the periphery of the text. Appearing to exist as the sole figure centrally located, the clothed body of Christ serves as the unifying principle or centrifugal force of the text; it is counterbalanced by the centripetal forces represented by the swirling vortex of naked humanity at the periphery of the text. While the Christ-figure is bathed in a glowing light representative of Edenic tranquility, the multiple mass of humanity appears darkened, inverted, and unstable. While the solitary figure of Christ exists in an eternal, unchanging world, the crowded mass of humanity exists within a world of never-ending change. Entering a world whose centripetal forces threaten to disrupt existence, Man and Woman maintain physical—and spiritual—connection to the centrifugal force emanating from the Christ-figure. Though they are on the path leading toward hellfire and damnation, they are also on the same path that leads to paradise and salvation. The only hope of transcendence—to override or rise above the earthly world—comes by following the figure of Christ from the world of innocence to the world of experience.

Blake's religious imagery in the *Fall of Man* may appear far removed from modern scientific belief, the subject of this work. It may appear to be especially far removed from modern evolutionary science, which has arguably offered an account of human origins that has substituted (if not subverted outright) the Biblical narrative represented in Blake's *Fall of Man*. Yet many of the themes in Blake, I would argue, run equally deep within the narrative of the history of science in the West.[4] Among these are the need to reconcile, to bring to line, to unify within a single, all-embracing, coherent, and logical system of thought those divergent—and diverse—elements that threaten to disrupt an orderly world. From Heraclitus, who sought the one in many, to Plato, who cherished the unity of knowledge, to the Enlightenment philosophers who sought to unify the branches of knowledge within a systematic and universal scheme, to the generations of positivists who dreamt of unifying the sciences, the narrative of the intellectual history of the West includes tales of heroic figures seeking unity in diversity, eternity within impermanence, and order in disorder. Solitary figures seeking the meaning of life, such

---

[4] Discussion of these themes, especially in the context of post-Enlightenment Western thought, has formed the corpus of Isaiah Berlin's work. See, for instance, Isaiah Berlin, *The Hedgehog and the Fox: An Essay on Tolstoy's View of History* (New York: Simon and Schuster, 1977, orig. pub. 1953); and the superb recent volume of his essays, *The Crooked Timber of Humanity: Chapters in the History of Ideas*, ed. Henry Hardy (New York: Vintage, 1992).

individuals also came together in collectives, building communities on common ground, sharing a belief in and a search for transcendent truths. Though traditionally set apart, and frequently seen at odds with each other, myth, religion, and science all share the same quest for universal, absolute, transcendent truths.[5] All three share a common opposition to views that deny the existence of transcendent truths in favor of local, relative, or otherwise "embodied," "contextual" forms of knowledge.

A similar tension between opposing points of view at the center versus the periphery of the text has made its way into contemporary intellectual circles via the introduction of multiculturalist sociopolitical theory. Using various devices to subvert, defuse, or deconstruct existing power structures inherent in any universalizing or totalizing systems, the goal of the multiculturalist project is to "give voice" to the diversity of positions silenced or marginalized by moves toward unity. While it appears an ideological political action program, multiculturalism also rests tremulously on epistemic foundations that hold that knowledge is culturally, historically, and "locally" constructed, a position antithetical to universal or totalizing systems of knowledge that attempt to transcend both history and culture. Thus, rather than upholding fixed, universal, essentialistic, or absolute truths, these multicultural theories of knowledge stress localized, contextual, and embodied features of knowledge (thus the oft-heard phrase that knowledge is a cultural or social construct); as such, they run contra to those Western intellectual traditions like science that hold to some notion of transcendent truth (removed from history and culture) and that, by historical definition, attempt to order and systematize knowledge of the world.

At present, multicultural theories of knowledge, in various guises (included here are postpositivist, post-Enlightenment theories of knowledge),[6] are making their way into the study of science, arguing among other things against the unity of knowledge, against coherent abstract logical principles, arguing instead for localized, contextual, and disunified practices. Whereas the introduction of multicultural epistemic frame-

[5] For a recent discussion, see Robert M. Torrance, *The Spiritual Quest: Transcendence in Myth, Religion, and Science* (Berkeley: University of California Press, 1994). Gerald Holton alludes to a similar "subterranean link between science and religion" on p. 135 in an essay entitled "The Controversy over the End of Science." See Gerald Holton, *Science and Anti-Science* (Cambridge, Mass.: Harvard University Press, 1994). See also Paul Forman, "Independence, Not Transcendence, for the Historian of Science," *Isis* 82 (1991): 71–86.

[6] These terms are frequently used in tandem with postmodernism and poststructuralism. Though they are frequently used synonymously, they bear different meanings for different academic communities within the humanities. Because I wish not to burden the reader with discussion of the shades of meaning in these terms (this would be a heroic effort in itself), I have instead chosen the terms that bear most meaning to historians and philosophers and to the present subject of study.

works has generated some of the most intellectually exciting scholarship in the social sciences and humanities, it has also garnered negative attention from self-appointed science watchdogs like Paul R. Gross and Norman Levitt, who, in their recent polemic *Higher Superstition: The Academic Left and Its Quarrels with Science* have derided (and not without some justification) much of the literature that has made its way into science studies.[7]

The present work offers a narrative that is situated within both of these contrasting positions. It draws heavily from science studies and cultural studies, the true multicultural bête noire, of Gross and Levitt's *Higher Superstition*. But while it borrows from some of the most provocative of this literature within the humanities (I refer here to cultural history/cultural study, literary theory, and philosophy of history), it does so to narrate a story that has some meaning for the community of scientists historically and intellectually involved in the project. Whereas the narrative tells of the heroic struggle to unify the branches of knowledge within a positivist theory of knowledge from what students of culture call, sometimes unhappily, an "insider's" perspective, the story is only possible because it simultaneously embraces a multicultural, postpositivist perspective that gives it enough of a critical distance to make it also an "outsider's" perspective. Taking the perspective of the scientist or enculturated member through a 359-degree turn, it tries to recover a critical vantage point from which to observe a historical event of some importance. It also attempts to recover what some students of culture have held in disrepute, the genre of the grand or unifying narrative. As the project tries to capture the perspective of the scientists studied herein, all historical measures are taken to retell a story that has meaning to the participants, all of whom searched for coherent, unifying theories. In keeping with some cultural and science studies movements that envision science as a culture, the language, rituals, practices, and cosmologies of the members are also considered in the project. Science is thus "contextualized" as traditional "external" and "internal" are collapsed within a historical perspective that recovers the "language" of science. Scientific content—lost in many of the recent historical attempts to understand science—is an integral component of this story.

Among other things, the present work tells a story of the moment in

---

[7] Paul R. Gross and Norman Levitt, *Higher Superstition: The Academic Left and Its Quarrels with Science* (Baltimore: Johns Hopkins University Press, 1994). Less polemical analyses of science and its critics can be found in Lewis Wolpert, *The Unnatural Nature of Science* (Cambridge, Mass.: Harvard University Press, 1992); and Holton, *Science and Anti-Science.*

the intellectual history of the West (one of the cultural categories operant here) when diverging points of view appeared to converge within a unified logical system of thought. It tells a story of a historical event that appeared to fulfill a project at least as deep as the Enlightenment project (or even deeper still) of unifying the branches of knowledge. It tells a story of the emergence, unification, and maturation of the central science of life—biology—within the positivist ordering of knowledge; and it tells a story of the emergence of the central unifying discipline of evolutionary biology (complete with textbooks, rituals, problems, a discursive community, and a collective historical memory to delineate its boundaries). Sustained by a linkage of the autonomous disciplines of knowledge, the proper systematic study of "Man"—his origins and location within a progressive cosmological scheme—became cojoined and reducible through logic to the mechanistic and materialist frameworks of the physical sciences. What effectively emerged was an evolutionary worldview, a cosmology, and a poetic weltanschauung, fulfilling an intellectual project that began with the very origins of the narrative of science in Western culture.[8]

The historical event in question took place between the third and fourth decades of the twentieth century during the interwar period. It is recognized by various terms, the most appropriate of which for historians is the "evolutionary synthesis." Variously—and confusingly—interpreted by evolutionary biologists, historians, and philosophers to the point of engendering some of the most acrimonious debates in contemporary evolutionary biology, it has remained a cornerstone of the history of modern evolutionary thought, grounding contemporary biological theories ranging from sociobiology to exobiology. The story told herein recognizes the contributions of the evolutionary synthesis, and attempts to understand why it has occupied the thoughts of biologists, historians, and philosophers of science. As the story unfolds it also tells of the origin of a group of "architects" and "unifiers," members of a unifying discipline called evolutionary biology that in turn would form the unifying principle of the modern biological sciences, serving ultimately as the fulcrum of a liberal, progressive, humanistic, and evolutionary worldview.

Although the narrative of *Unifying Biology* should ideally speak meaningfully to the community of scientists involved in its formation, without prolonged theoretical or methodological discussion, failure to include these features would disengage potential audiences from the humanities.

---

[8] For a clear demonstration of this poetic weltanschauung, see Connie Barlow, ed., *Evolution Extended: Biological Debates on the Meaning of Life* (Cambridge: MIT Press, 1994).

Because the methodology or historiography (literally the writing of history) may be of equal interest to the humanities as the narrative proper, the book is organized so that it will facilitate enough comprehension to provoke discussion across audiences from the sciences to the humanities. Readers might engage relevant parts selectively as they wish. Part 2 (including the narrative of *Unifying Biology: The Evolutionary Synthesis and Evolutionary Biology*, which has already appeared in print in a somewhat shorter version) can stand alone, or it can be read along with the other parts.[9]

In addition to chapter 1, the present exegetical opening, which hints at the broader epistemic, political, and existential issues raised by *Unifying Biology*, I have included several other chapters to set the stage for the narrative for varied audiences. All of these are assembled in part 1, entitled "History, Theory, and Practice." Readers who are not familiar with the evolutionary synthesis, its importance, and the mass of confusing literature on the subject may benefit by reading chapter 2, "A 'Moving Target': Historical Background on the Evolutionary Synthesis," which includes a review of relevant literature as it traces developments in recent evolution through to the 1980s. Historical and analytical discussion is continued in chapter 3, "Rethinking the Evolutionary Synthesis: Historiographic Questions and Perspectives Explored," which poses questions of existing approaches, explores alternative modes of historiographic inquiry, and criticizes possible story lines. Chapter 4, entitled "The New Contextualism: Science as Discourse and Culture," takes readers historically through much of the recent literature in science studies and cultural studies and develops a means for contextualizing the evolutionary synthesis. Among other things, it engages in discussion of the form of contextualist historiography developed for *Unifying Biology*, so that the theoretical and methodological scaffolding for its narrative construction is revealed. Scientific readers may wish to skip this section, though I make an effort to engage their readership as best I can.

I have also added chapter 6 entitled "Reproblematizing the Evolutionary Synthesis" in part 3 ("Persistent Problems"), which breaks with the constraining narrative format of chapter 5 to respond directly to the potential concerns of readers, to add additional clarification, and to provide pointers for further study into the history of evolution and biology. The "Epilogue" returns to the deeper undergirding themes of *Unifying Biology*

[9] V. B. Smocovitis, "Unifying Biology: The Evolutionary Synthesis and Evolutionary Biology," *Journal of the History of Biology* 25 (1992): 1–65.

introduced in the exegetical opening. Although it was not my original intention, the present organization does bear close resemblance to the standard scientific paper: abstract, introduction, materials and methods, results, and discussion. Very probably, this is a carryover from a problem-oriented view of history for which existing tools and instruments (in the way of historiography and cultural theory) have been developed and applied. In the conventional manner of the scientific paper, the discussion and conclusions address remaining problems and point to possible areas for further inquiry. This alternative way of envisioning the organization in its entirety may aid some readers, who may be perplexed by the content, tone, and organization of the chapters.

WRITING *Unifying Biology*

> The grand narrative has lost its credibility, regardless of what mode of unification it uses, regardless of whether it is a speculative narrative or a narrative of emancipation.
>
> Jean-François Lyotard, *The Post modern Condition*

One would be hard-pressed to envision both the history of modern science and extant scientific practice—especially for historical sciences like evolution—without belief in grand narratives, or universalizing stories about the world. Reports of the death, demise, or "loss of credibility" of grand narrative are thus somewhat exaggerated, very possibly limited to those individuals living at the periphery of the text within their "Postmodern Condition" or those otherwise having little contact with historical sciences.[10]

---

[10] See Jean-François Lyotard, *The Postmodern Condition: A Report on Knowledge*, trans. Geoff Bennington and Brian Massumi (Minneapolis: University of Minnesota Press, 1988). For a modification, explication, and use of Lyotard's "grand narrative," see Alan Megill, " 'Grand Narrative' and the Discipline of History," in Frank Ankersmit and Hans Kellner, eds., *A New Philosophy of History* (Chicago: University of Chicago Press, 1995), pp. 151–73. Megill uses the phrase to mean a story (arranged with beginning-middle-end) that is universalizing or all-embracing as distinct from "master narrative," which is a segment of grand narrative usually concerning the nation or state. Drawing on Lyotard for his definition of grand narrative, he writes: "I intend the term more broadly, to designate a vision of coherence—in particular, a vision of coherence broad enough to support objectivity claims" (p. 264). My use echoes Megill's and includes the originary history of life, the universe, and humanity. For a recent analysis of the use of grand narrative (and master narratives) in the writing of American history, see Dorothy Ross, "Grand Narrative in American Historical Writing: From Romance to Uncertainty," *American Historical Review* 100 (1995): 651–77. For an introduction and discussion of some well-known social theorists (including Thomas

Because the narrative construction is one of the critical features of *Unifying Biology*, special attention may be paid to its design, beyond that ordinarily paid to written works. The title is in the present-participle-plus-object grammatical form, referring to the ongoing *process* of scientific knowledge making and the ongoing process of *unifying biology*. The first portion of the subtitle refers to the historical event of the evolutionary synthesis, and alludes to the manner in which the historical event of the synthesis contributes to this process, as well to the manner by which consensually derived historical events serve to support the formation of a collective through a shared historical memory (in this case, a disciplinary collective). That a collective memory emerges concomitant with the scientific discipline (complete with texts, problems, a community, and other cultural practices) is indicated by the second portion of the subtitle, evolutionary biology, the disciplinary category lending coherence to the community of scientists involved. Taken as a whole (and read backward) the title summarizes the story line that recounts the emergence of a unifying discipline of evolutionary biology during the period of the evolutionary synthesis, all of which aid the process of unifying biology.

Close attention should be paid to the initial formulation of the problem of the synthesis, the argument proposed, and the abrupt beginning of the narrative proper (in the way of a narrative "flashback"). The closing of the narrative proper begins with the arrival of the unified science of biology (or what appears to be the unified science of biology) in approximately 1955. The section entitled "Postscript" brings the reader up to the recent present. This carries the narrative of the synthesis through to the next generation, demonstrating the genealogical continuity—and discontinuity—in some of the most recent debates in evolutionary biology. This final section is designated as a postscript that is not intended as a detailed narrative of postsynthesis developments but merely as a demonstration of the continuity/discontinuity in the narrative line.

The final paragraph begins to close the narrative of the evolutionary synthesis where the problem was articulated and subsequently disseminated as a "historical event"—Mayr and Provine's 1980 book (which also effectively served as a textbook for future historians of the synthesis). In so doing, the final paragraph thus situates the author within the larger narrative line. Thus, the personal narrative is effectively rewoven within

---

Kuhn, Michel Foucault, and Jacques Derrida) who have challenged the existence of universalizing systems of knowledge, the positivist ideal of the unification of the sciences, and the legitimacy of grand theory and narrative, see Quentin Skinner, ed., *The Return of Grand Theory in the Human Sciences* (Cambridge: Cambridge University Press, 1985).

the narrative of the synthesis, all against the contextual backdrop of "Western thought." In the language of literature, I have written myself into this story, for as an enculturated member in the collective—an "insider"—it is also *my* story, and I have indicated so by placing my own script or signature (in the way of signing off) within the narrative. Because I do not wish to burden audiences with self-reflexive historical musings, I have truncated the text.[11] It may appear to some as being overly cryptic, but this is not my aim. The cyclical and genealogical pattern in the narrative may also be apparent as the beginning and end dovetail into each other, or in the language of literature, *arche* and *telos* are one. The historical framework thus appears to be not only presentist, but epistrophic, in that it also bears a cyclical pattern of return. As phenomenologist David Carr eloquently states in the introduction to his *Time, Narrative, and History*, "the present study, though it hardly qualifies as a story, illustrates one of the most important features of lived time, narrative, and history itself that we shall be discovering along the way: namely, that only from the perspective of the end do the beginning and the middle make sense."[12] *Unifying Biology* may qualify as just such a story.

The meter, choice of language, and selection of metaphors in the text are deliberately constructed so as to represent the making of scientific knowledge as poetic, expressive, or emotive activity. The language also attempts to capture the sense of heroic, ironic, and dramatic struggle in the reworking of an ancient mythic narrative of human and cosmic origins. The language is the language of the architects as represented in their texts, both public and private, followed by the author's genealogical "voice" (see the discussion on situating the author above and below). Rather than rely on militaristic metaphors, economic metaphors (which include the language of production/consumption), or the language of self-interested "pop psychology," all of which represent a culture of science that is unsatisfying, the language used is humanistic and aesthetic and draws on scientific knowledge as emotional expression. Science, in this view, is an expression of a desire to understand and formulate a meaningful worldview within the Western way of thought. The text is consequently full of metaphors that attempt to represent the perspective of the scientists engaged in their project. In the barest of aesthetic terms,

---

[11] For further discussion on history, memory, narrative, and the self, see Mark Freeman, *Rewriting the Self: History, Memory, Narrative* (New York: Routledge, 1993).

[12] David Carr, *Time, Narrative, and History* (Bloomington: Indiana University Press, 1986), p. 6.

the historical actors are represented as emotive creatures experiencing desires, beliefs, and hopes; individuals within collectives striving to communicate with others through talk and dialogue; storytellers searching for the meaning of life; sense makers and worldview builders, struggling to find a balance between opposing points of view. At a deeper level, this is part of an "aesthetic" or "emotive" epistemology.[13]

No doubt, some readers will be taken aback—if not annoyed—by the combination of the extensive use of lengthy footnotes, the construction of the narrative, and the meter of the prose. These features of the text are not fortuitous, nor do they reflect the author's idiosyncratic writing style, but they emerge from the design of the narratological framework, which attempts to reach diverse audiences. In following the genre of the "grand unifying narrative" many of the details, asides, and digressive strands of the history have been rewoven so as to "flatten out" or streamline the history into a progressive causal narrative of the rise of the unifying discipline of evolutionary biology. To preserve the narrative drive of the text (as set up critically in the introductory section), I have therefore concentrated on following through the story line to its completion, and have relied on the footnotes as a means of including clarifying statements to readers, in addition to recognizing and responding to pertinent literature, both primary and secondary. Although several widely accepted styles of historical writing reject the use of footnotes as explanatory devices, points of clarification, additional asides, and the like (historians are often heard to say that if you can't put the thought in the main text, don't put it in at all), other genres of historical writing, especially common in intellectual history, critically rely on the heavy use of footnotes to explain critical points in the text. For examples of this genre of historical writing, readers may wish to consult the *Journal of the History of Ideas*, which still endorses the heavy use of explanatory footnotes where need be. Though it is not generally recommended as an exemplar for contemporary writers, John Theodore Merz's *A History of European Thought in the Nineteenth Century* (Edinburgh: W. B. Blackwood, 1903–14) still remains a fascinating text for readers because of its extensive and richly woven messages embedded in the footnotes.

The verb tenses used in *Unifying Biology* have also been deliberately chosen to reflect a processual or historical coming to be (or becoming).

---

[13] My position here echoes some of Susanne K. Langer's philosophy and theory of art. See her corpus of work, but especially *Philosophy in a New Key* and its elaboration in *Feeling and Form*: Susanne K. Langer, *Philosophy in a New Key: A Study in the Symbolism of Reason, Rite, and Art*, 3d ed. (Cambridge, Mass.: Harvard University Press, 1957); and idem, *Feeling and Form: A Theory of Art* (New York: Charles Scribner's Sons, 1953).

Rather than the simple past tense, which is almost of formulaic use for historians, the narrative relies heavily on variations of the past tense that can help transmit continuous movement in time. Although contemporary writing also calls for use of the active voice (some historians recommend abolishing the passive voice in historical prose entirely), *Unifying Biology* instead relies heavily on the use of the passive voice because it narrates the history of a discipline. Thus, rather than imparting too much action and agency to any one individual, the text imparts action to the narrative script that is driving the historical actors and their science to performance.

Multiple readings of *Unifying Biology* may be useful: the first to experience the narrative "drive" of the story, the meter of the prose, and the progression of the sequences (designated by subheadings) that make up the story line; a second to peruse the asides and explanatory or clarifying points in the footnotes; a third to assimilate properly all this material within the historiographic concerns raised in the preceding parts.

Because, too, it offers what we may view as a framework for further interpretation (in what we may call an interpretive framework) for the history of modern evolution and biology, *Unifying Biology* may also be considered a "synoptic" historical account. By synoptic I do not mean a summary (in the rather reductionist sense of summation), but the attempt to bring together, visually or optically, multiple objects within a single field of historical vision. Objects far apart in one view may appear close together in another, or may appear in direct alignment, depending on their relation to each other and to the point of view of the observer. In the language familiar to microscopists, *Unifying Biology* observes a historical "object" with low-magnification lenses. Readers who wish to image what a higher-magnification (or higher-resolution) historical account looks like within the same interpretive framework may consult my "Organizing Evolution: Founding the Society for the Study of Evolution (1939–1950)" (*Journal of the History of Biology* 27 [1994]: 241–309), which offers a historical reconstruction of a more "local" feature of the synthesis based on documents deposited by successive secretaries of the Society for the Study of Evolution. Though the "final" stories differ somewhat in their emphases, both are still focused on the same historical "object" of study. It should also be kept in mind that one goal of the project is to give "form" to a surficial—rather than superficial—historical "object" of study that has escaped all attempts at understanding thus far (the "moving target" of chapter 2).

Similarly, *Unifying Biology* may also be viewed as a "syncretic" account in that it tries to reconcile diverging or different points of view within a unifying narrative. One result of these approaches is that the movement toward the unification of evolution and biology is imaged alongside what may be viewed as "parallel" philosophical movements associated with positivism. By positivism, I mean that philosophical movement that grew out of the Enlightenment most closely associated with the work of Auguste Comte. Later filiations of positivism led to the philosophy of Ernst Mach, and later still manifestations of positivism became wedded to mathematical logic to found the school of logical positivists of the Vienna Circle (who later altered their identities to become logical empiricists). Definitional criteria of "positivism" in its varied historical guises, currents, streams, or movements have been attempted, though it should be noted not without great success in varied works of reference.[14] Here it is applied to as an overarching—or undergirding—epistemic packaging of values, beliefs, and practices that upheld antimetaphysical, analytical, scientific—and scientistic—notions about proper methodology, as well as formalized relations among the sciences. All forms of positivism sought ultimately a unified theory of knowledge. While the architects of the evolutionary synthesis were not openly "influenced" by their positivist contemporaries, they and their positivist cohorts had shared inherited commitments and assumptions about proper scientific methodology.[15] Philosophers like Susanne Langer, other intellectual historians, and historians of philosophy, science, and art, as well as historians associated with the "new" cultural history, have attempted to deal with issues foregrounded in terms such as "mentalities" to describe what I term as "epistemic framework" or "packaging."[16] Other terms echo these, such as worldview, framework, discursive mentalité, cosmologies, narrative worlds, or weltanschauung. I am here designating the positivist theory of

[14] For brief explication, see the entries under "Positivism" and "Logical Positivism" in Paul Edwards, ed., *The Encyclopedia of Philosophy* (New York: Collier Macmillan, 1967). See also Alfred J. Ayer, ed., *Logical Positivism* (Glencoe, Ill.: Free Press, 1959).

[15] I am not here engaging in a comparative analysis along the lines of Laurence D. Smith's study of the effects of logical positivism on the behavioral sciences. See Laurence D. Smith, *Behaviorism and Logical Positivism: A Reassessment of the Alliance* (Stanford: Stanford University Press, 1986).

[16] See n. 13. An excellent abbreviated introduction is Patrick H. Hutton, "The History of Mentalities: The New Map of Cultural History," *History and Theory* 20 (1981): 237–59; and see also idem, *History as an Art of Memory* (Burlington: University of Vermont, 1993). For recent and more familiar attempts to grapple with this issue in the context of the new cultural history, see Roger Chartier, *Cultural History: Between Practices and Representations*, trans. Lydia G. Cochrane (Ithaca: Cornell University Press, 1988); and Lynn Hunt, ed., *The New Cultural History* (Berkeley: University of California Press, 1989). See also the collection of essays in Michel Vovelle, *Ideologies and Mentalities*, trans. Eamon O'Flaherty (Chicago: University of Chicago Press, 1990).

knowledge as the legitimating background of inherited values (many of which are silent) against which evolution and biology would emerge as legitimate sciences. This is in direct response to interpretations that would legitimate science in terms of immediate social "interests." Reasons for the introduction of the positivist dimension are included in chapters 3 and 6.

No doubt some will remain unconvinced of some of the causal connections that are drawn by the story line in the narrative herein. Readers here should note that my goal is not to argue that this is the final "straight story,"[17] but to release the text of *Unifying Biology* so that it can bring scholars from both the sciences and the humanities into lively discussion. It is my hope that it will invite further discussion of not only the interpretive features of the evolutionary synthesis—an event of critical importance to the history of scientific knowledge—but also to draw attention to the potential of cultural, narrative, literary, and/or discursive study of scientific knowledge.[18] Specifically, it is hoped that this study will explore the narrative pattern of historical sciences like evolution,[19]

[17] See Hans Kellner, *Language and Historical Representation: Getting the Story Crooked* (Madison: University of Wisconsin Press, 1989). The most recent account of the problems of historical writing are discussed in Frank Ankersmit and Hans Kellner, eds., *A New Philosophy of History* (Chicago: University of Chicago Press, 1995; orig. pub. London: Reaktion Books, 1995). On the aims and problems of historical writing, see also the classic corpus of work by Hayden White. See especially his essay, "The Fictions of Factual Representation," in Hayden White, *Tropics of Discourse: Essays in Cultural Criticism* (Baltimore: Johns Hopkins University Press, 1978). See also Hayden White, *The Content of the Form: Narrative Discourse and Historical Representation* (Baltimore: Johns Hopkins University Press, 1987). The body of work on the literary aspects of historical writing is vast; but see especially the classic literature that explores historical writing and knowledge as narration: Arthur C. Danto, *Narration and Knowledge* (New York: Columbia University Press, 1985); and Louis Mink, "Narrative Form as a Cognitive Instrument," in R. H. Canary and H. Kozicki, eds., *The Writing of History: Literary Form and Historical Understanding* (Madison: University of Wisconsin Press, 1978).

[18] The literature exploring science as literary, discursive, or textual activity is growing rapidly but varies enormously in its aims and theoretical scaffolding. Among general works, see the diversity of approaches in Charles Bazerman, *Shaping Written Knowledge: The Genre and Activity of the Experimental Article in Science* (Madison: University of Wisconsin Press, 1988); Peter Dear, ed., *The Literary Structure of Scientific Argument: Historical Studies* (Philadelphia: University of Pennsylvania Press, 1991); David Locke, *Science as Writing* (New Haven: Yale University Press, 1992); and Marcello Pera, *The Discourses of Science*, trans. Clarissa Botsford (Chicago: University of Chicago Press, 1994; orig. pub. as *Scienza e retorica*, by Gius. Laterza and Figli, 1991). More closely compatible with *Unifying Biology* is the first (spring) issue of *Science in Context* 7 (1994), which includes a suite of essays exploring the narrativity of science edited by Joseph Mali and Gabriel Motzkin. Two additional authors have just completed or are presently preparing book-length treatments exploring the narrative patterns in science or science as discursive activity: Joseph Rouse, *Engaging Science* (Ithaca: Cornell University Press, 1996); and Alison Wylie, *No Return to Innocence* (Princeton: Princeton University Press, forthcoming).

[19] Study of the narrative features of evolution has not gone without notice. See Thomas Goudge, *The Ascent of Life: A Philosophical Study of the Theory of Evolution* (Toronto: Univer-

the role that narratives play in constructing and delineating the boundaries of disciplines like evolutionary biology, how such narratives help construct the disciplinary memories and identities of members, as well as the even more ambitious goal of opening discussion on the meaning of causality, temporality, and narrativity in both history and science.[20]

Throughout this work, I have tried to keep the prose spare and free of abstruse and arcane terms insofar as possible, although I realize that given the diverse intellectual communities and disciplines that it speaks to on technical points (in both evolutionary science and contemporary cultural theory), this is a practically impossible goal. Despite the brevity and sparseness of the following text, I have not taken the profound nature of this project lightly. At stake in the subject matter of *Unifying Biology* is not only the modern Neo-Darwinian synthesis of evolution and the cosmological apparatus or worldview of the "Enlightened West," but the very foundations of knowledge, the range of choices of sociopolitical programs of action, and the meaning of life itself: such is the power of *this* grand narrative that it exposes the epistemic, political, and existential angst of the late twentieth century.

Admittedly, the goal of retelling this grand a story in a manner that takes diverse perspectives into account is so hopelessly ambitious that it will probably fall woefully short of its mark. Yet the urge to retell it arises from a need so compelling that the story, which takes on a life of its own, practically tells itself. Like a script that runs itself through the historical actors in a dramatic play, so that they are moved to performance in an externally determined plot line, the narrative running through *Unifying Biology* can be seen as so grand and totalizing that it writes itself through author/actors cast by their history and culture to play those roles.

---

sity of Toronto Press, 1961); David L. Hull, *The Metaphysics of Evolution* (Albany: State University of New York Press, 1989); Greg Myers, *Writing Biology: Texts in the Social Construction of Scientific Knowledge* (Madison: University of Wisconsin Press, 1990); Misia Landau, *Narratives of Human Evolution* (New Haven: Yale University Press, 1991); Robert J. O'hara, "Telling the Tree: Narrative Representation and the Study of Evolutionary History," *Biology and Philosophy* 7 (1992): 135–60. For an especially lucid discussion on the narrative structure of evolution and an interesting account of "triangulation," see Robert J. Richards, "The Structure of Narrative Explanation in History and Biology," in Matthew H. Nitecki and Doris V. Nitecki, eds., *History and Evolution* (Albany: State University of New York Press, 1992), pp. 19–53; and see also David L. Hull's contribution to the same volume, *History and Evolution*, entitled "The Particular-Circumstance Model of Scientific Explanation," pp. 69–80.

[20] See Carr, *Time, Narrative, and History*; and see the corpus of work by Paul Ricoeur, *Time and Narrative* (Chicago: University of Chicago Press, 1988; orig. pub. as *Temps et Récit*, by Editions Du Seuil, 1983).

PART ONE

History, Theory, and Practice

CHAPTER TWO

A "Moving Target": Historical Background
on the Evolutionary Synthesis

> The evolutionary synthesis is unquestionably an event of first-
> rank importance in the history of biology.
>
> William B. Provine, *The Evolutionary Synthesis*

THE EVOLUTIONARY SYNTHESIS

From Julian Huxley (grandson of Thomas Henry Huxley, Darwin's
famed "Bulldog"), who first used the word "synthesis" in 1942 to describe the newer and "modern synthesis" of evolutionary thought,[1] to
contemporary biologists, historians, and philosophers of science, the
sense has existed that an intellectual event of great importance took place
during the interwar period. So important is the event that it has drawn
the attention of scores of individuals who have studied it for a variety of
reasons and from many vantage points over the years. Despite energetic
efforts to understand what constituted this event, no consensus has yet
emerged as to its exact nature. Instead of agreement on even the most
fundamental of points, scholars have only generated disagreement and/or
downright discord. As the numbers of individuals turning their attention
to the synthesis has grown, so too has the number of conflicting and
contradictory accounts. Gallons of ink have been spilt on the subject, and
heated exchanges have taken place. Within the evolutionary community—no stranger to contentious issues—the subject of the proper interpretation of the synthesis has attained an unsurpassed historical notoriety. Because it has only generated more and more confusion and
contradiction with time, historians of biology generally consider it one of
the central problems of the history of biology.

THE EVOLUTION OF THE SYNTHESIS

One way to sort out the problems associated with this complex episode is to examine the historical unfolding—or evolution—of the syn-

---

[1] Julian S. Huxley, *Evolution: The Modern Synthesis* (London: Allen and Unwin, 1942).

thesis itself. Although the term "synthesis" holds a range of meanings for contemporary evolutionists, ranging from the theory of evolution formulated during the 1930s and 1940s, to the historical event associated with that interval of time, to the convergence between relevant biological disciplines, it was best articulated by Julian Huxley, who first coined the phrase.

The account originally provided by Huxley held that developments in the new science of genetics (which had emerged as a scientific discipline with the rediscovery of Mendel in 1900) and developments in selection theory (largely as the result of mathematical and experimental modeling studies on evolution) had made possible a "synthesis" between what had previously been thought contrasting or irreconcilable points of view in biology in the 1920s.[2] The same synthesis, which involved the application of experimental and quantitative methods to evolution and biology, also led them to become more rigorous sciences that could finally rival exemplar sciences like physics and chemistry. Because it involved a fusion between the newer genetics and the older Darwinian selection theory, which had previously represented different and fragmented portions of biology, the new synthesis also unified several branches of the biological sciences so that they would form a unified science of biology. Through this unification, Darwin's controversial "mechanism" of natural selection emerged as the primary cause of evolutionary change. Darwinism itself, which had been "eclipsed" (in Huxley's well-known phrase) at the turn of the century by rival theories of evolution, was raised to the status of not only a palatable but also a preeminent scientific theory. The end result was an effectively "modernized" evolutionary theory that was commensurate with developments in recent sciences like genetics and what Huxley had previously termed his "new" systematics.[3]

Huxley's own call to recognize the modern synthesis of evolution appeared shortly after the first synthetic account of evolution had been written by Theodosius Dobzhansky, a Russian émigré. Appearing in 1937, his *Genetics and the Origin of Species*[4] was the first successful attempt to synthesize theoretical studies of evolution that had emerged from mathematical models by Sewall Wright in the United States (and in Britain by R. A. Fisher and J. B. S. Haldane) with natural, populational

---

[2] See ibid. for the original summary.

[3] Julian Huxley, *The New Systematics* (Oxford: Oxford University Press, 1940). This book had described the most recent developments in systematics that had incorporated populational approaches with genetic and ecological studies.

[4] Theodosius Dobzhansky, *Genetics and the Origin of Species* (New York: Columbia University Press, 1937).

studies made by systematist-naturalists and field biologists like Dob-zhansky (and in Britain by Fisher's co-worker, E. B. Ford). The book was not only widely read, but rapidly began to function as a textbook for younger evolutionists who were attracted to the more dynamic new study of evolution in the late 1930s. The book, and the charismatic Dob-zhansky, catalyzed the publication of additional evolutionary works, which lent support for the belief that a new synthesis of evolution had occurred or was actively occurring. A series of books and articles by biol-ogists representing diverse biological disciplines appeared in the decade between 1940 to 1950 that echoed Dobzhansky and each other. These included (in chronological order) C. D. Darlington, *Evolution of Genetic Systems* (1939), Julian Huxley, ed., *The New Systematics* (1940), Ernst Mayr, *Systematics and the Origin of Species* (1942), G. G. Simpson, *Tempo and Mode in Evolution* (1944), M. J. D. White, *Animal Cytology and Evo-lution* (1945), Bernhard Rensch, *Neuere Probleme der Abstammungslehre* (1947), Glenn L. Jepsen, Ernst Mayr, and G. G. Simpson, eds., *Genetics, Paleontology and Evolution* (1949), and G. Ledyard Stebbins, Jr., *Varia-tion and Evolution in Plants* (1950).[5] Each played a critical—though dif-ferent—role in contributing to the ferment of intellectual activity sur-rounding evolution at this time.

With the end of the Second World War (which had impeded oppor-tunities for gatherings, conferences, and publications, in addition to non-war related research), evolutionists (who included the individuals noted above) recognized the "modern" synthesis of evolution, the end of a pe-riod of dissonance, and the convergence of biological disciplines by cele-brating officially at a special conference gathering in Princeton in 1947.[6] By that time, too, the same group had formed an international journal-

[5] A complete list of the central texts should ideally include the published work of the theoretical population geneticists (in chrononological order): Ronald A. Fisher, *Genetical Theory of Natural Selection* (Oxford: Oxford University Press, 1930); Sewall Wright, "Evo-lution in Mendelian Populations," *Genetics* 16 (1931): 97–159; E. B. Ford, *Mendelism and Evolution* (London: Methuen, 1931); and J. B. S. Haldane, *Causes of Evolution* (London: Longmans, Green, 1932). Some commentators include Julian S. Huxley, A. C. Hardy, and E .B. Ford, eds., *Evolution as a Process* (London: Allen and Unwin, 1954), as one of the texts, though it is more commonly held that G. Ledyard Stebbins, *Variation and Evolution in Plants* (New York: Columbia University Press, 1950), closed the evolutionary synthesis. The complete works of the evolutionary synthesis are included in Ernst Mayr and William B. Provine, eds., *The Evolutionary Synthesis: Perspectives on the Unification of Biology* (Cam-bridge, Mass.: Harvard University Press, 1980).

[6] For the proceedings of the Princeton bicentennial conference, see Glenn L. Jepsen, Ernst Mayr, and George Gaylord Simpson, eds., *Genetics, Paleontology, and Evolution* (Princeton: Princeton University Press, 1949). A preliminary summary, the full program, conference participants, and a group photograph of the conference was published after the conference in 1946: see Glenn L. Jepsen and Kenneth Cooper, *Genetics, Paleontology, and Evolution* (Princeton: Princeton University Press, 1946).

issuing society for the study of evolution (the Society for the Study of Evolution [SSE]). The journal that would henceforth serve as the common forum for discussion of the new dynamic science of evolution was called, appropriately enough, *Evolution*.[7]

But while a growing audience engaged in celebrating the "rebirth of Darwinism" recognized that a synthesis of evolution was taking place, the meaning of Huxley's synthesis was altered and permuted as it was used by its many subsequent followers. In time biologists came to associate the interwar period and developments in evolution with the establishment of what they variously termed the synthetic theory of evolution, the Neo-Darwinian synthesis, Neo-Darwinism, and later the evolutionary synthesis, in addition to Huxley's original "modern synthesis of evolution." Few reflective biologists realized that these terms, which were used interchangeably, did not necessarily have synonymous meanings or refer to the same thing.

The confusion surrounding the significance of synthesis actually began as soon as attempts were made to assess its status. The first historically noteworthy attempt actually led to depressing results: rather than assess the contributions of specific fields, individuals, or developments, it only made apparent those biological fields that had not been represented at all. At an Oxford Symposium of the Society for the Study of Experimental Biology, embryologist Conrad H. Waddington raised a voice of dissent as he pointed out that whereas the synthesis was successful in areas like mathematical population genetics, it left out his own field of embryology.[8] Embryology, which had been a cornerstone of nineteenth-century biology and had contributed the very term "evolution" (meaning an unfolding) to Darwin's theory of descent with modification, had been a critical component of evolutionary theory; its absence indicated some potential problems with the new modern synthesis of evolution, especially with respect to embryology.[9]

[7] For a history of organizational efforts to found the SSE and the journal *Evolution*, and Ernst Mayr's role in these, see Vassiliki Betty Smocovitis, "Organizing Evolution: Founding the Society for the Study of Evolution (1939–1950)," *Journal of the History of Biology* 27 (1994): 241–309; see also "Disciplining Evolutionary Biology: Ernst Mayr and the Founding of the Society for the Study of Evolution, and *Evolution* (1939–1950)," *Evolution* 48 (1994): 1–8. For a history of organizational efforts drawing on the precursor society's documents (the Committee on Common Problems of Genetics, Paleontology, and Systematics), see Joseph Allen Cain, "Common Problems and Cooperative Solutions: Organizational Activity in Evolutionary Studies," *Isis* 84 (1993): 1–25.

[8] Conrad H. Waddington, "Epigenetics and Evolution," in *Symposia of the Society for Experimental Biology: Evolution*, vol. 7 (New York: Academic Press, 1953), pp. 186–99. For further discussion of Waddington and embryology, see the narrative of *Unifying Biology* and "Reproblematizing the Synthesis" in chap. 6.

[9] For discussion on Waddington and embryology, see Scott Gilbert, "Bearing Crosses: The Historiography of Genetics and Embryology" (forthcoming).

Despite this and other such criticisms, evolutionists continued to revel in their "modern" synthesis. In the late 1950s evolutionists celebrated their synthetic evolutionary science at the Darwin Centennial Celebration at the University of Chicago, in honor of the one hundredth anniversary of the publication of Darwin's *On the Origin of Species*.[10] The special anniversary was also an opportune time to transmit the same synthetic theory and the restoration of Darwinism to American high school teachers in the hope of keeping the increasingly hostile hoards of American fundamentalists at bay: "One Hundred Years without Darwinism Are Enough," Hermann J. Muller stated emphatically in the title of a paper directed to American high school teachers.[11] At the same time, historical reassessments of evolution and its impact on Western thought with dramatic titles like *The Death of Adam* began to make their way to scholarly scientific and historical audiences as did detailed scholarly biographies representing different views of the Darwin figure.[12] The community of evolutionists also began to burgeon,[13] as American science boomed in the wake of "Atomic Age" and post-*Sputnik* efforts to lead the global community in science. The late 1950s and early 1960s witnessed a surge in scientific activity in the United States. The biological sciences, which appeared to hold great promise within the global community, were slated for heavy support in both research and teaching.

Within this period of biological activity, the torch of the spirit of synthesis was held high by a segment of the growing community of geneticists who were becoming actively interested in their history. Efforts to recognize the synthesis (sans the criticism) were led by individuals like Theodosius Dobzhansky, Hermann J. Muller, and others closest to population genetics. As early as 1950, at the occasion of the fiftieth anniversary of the "rediscovery of Mendel" and the birth of modern genetics,

[10] Proceedings were published in three volumes: Sol Tax, ed., *Evolution after Darwin*, vols. 1–2 (Chicago: University of Chicago Press, 1960); Sol Tax and Charles Callender, eds., *Evolution after Darwin*, vol. 3 (Chicago: University of Chicago Press, 1960). The third volume included historical memorabilia of the conference in the way of photographs and transcripts of workshop discussions. For a history of efforts to organize the celebration, see Vassiliki Betty Smocovitis, "Celebrating Darwin" (forthcoming).

[11] Hermann J. Muller, "One Hundred Years without Darwinism Are Enough," *School Science and Mathematics* 59 (1959): 304–16.

[12] See John C. Greene, *The Death of Adam: Evolution and Its Impact on Western Thought* (Ames: Iowa State University Press, 1959); and idem, *Darwin and the Modern World-View* (Baton Rouge: Louisiana State University Press, 1961). See also the volume edited by A. Hunter Dupree on Asa Gray's views of Darwinism: Asa Gray, *Darwiniana: Essays and Reviews Pertaining to Darwinism*, ed. A. Hunter Dupree (Cambridge, Mass.: Harvard University Press, 1963). The most widely read biography was by Gertrude Himmelfarb, *Darwin and the Darwinian Revolution* (Garden City, N.Y.: Doubleday, 1959). A historian of nineteenth-century England, Himmelfarb viewed Darwin in less than flattering terms.

[13] Smocovitis, "Organizing Evolution."

geticists launched their historical reflections with a special conference held 11–14 September by the Genetics Society of America for the Golden Jubilee of Genetics at Ohio State University. The proceedings, edited by L. C. Dunn as *Genetics in the 20th Century*, provided one of the first vehicles for histories of genetics.[14] In this volume, Theodosius Dobzhansky had the opportunity to reflect on the history of his own synthesis of evolutionary genetics.[15] His emphasis was, understandably, on the contributions of mathematical population genetics to the development of his population and evolutionary genetics. Later, at his 1955 address to the Cold Spring Harbor Symposium on Quantitative Biology on the subject of population genetics, Dobzhansky reinforced this sense.[16] The centrality of mathematical population genetics and the importance of population genetics to the newer synthesis was thus established by those geticists who turned to historical reflections on their work.

This stress on the contributions of the mathematical modelers and theoretical population geticists (like Haldane, Fisher, and Wright), and arguments for the centrality of mathematical population genetics in the synthetic theory quickly generated historical disagreement, especially from systematist-naturalists like Ernst Mayr. Inspired by Waddington's dissent,[17] Mayr raised his own objections to the growing historical consensus that theoretical genetics had played the greatest role in leading to the new synthesis of evolution. To Mayr, the synthesis was part of a

[14] L. C. Dunn, ed., *Genetics in the 20th Century: Essays on the Progress of Genetics During Its First 50 Years* (New York: Macmillan, 1951).

[15] Theodosius Dobzhansky, "Mendelian Populations and Their Evolution," in L. C. Dunn, ed., *Genetics in the 20th Century: Essays on the Progress of Genetics During Its First 50 Years* (New York: Macmillan, 1951), pp. 573–89. This had also been Dobzhansky's presidential address to the American Society of Naturalists.

[16] Theodosius Dobzhansky, "A Review of Some Fundamental Concepts and Problems of Population Genetics," *Cold Spring Harbor Symposia on Quantitative Biology* 20 (1955): 1–15. He wrote:

> The foundations of population genetics were laid chiefly by mathematical deduction from basic premises contained in the works of Mendel and Morgan and their followers. Haldane, Wright, and Fisher are the pioneers of population genetics whose main research equipment was paper and ink rather than microscopes, experimental fields, *Drosophila* bottles, or mouse cages. Theirs is theoretical biology at its best, and it has provided a guiding light for rigorous quantitative experimentation and observation. Experimental work in population genetics, initiated by Chetverikov some 30 years ago, has been growing apace. A body of evidence is now in existence to test the validity of some of the mathematical deductions. It is probably fair to say that accumulation of observational and experimental data is the order of the day. (pp. 13–14).

The symposium volume was entitled *Population Genetics: The Nature and Causes of Genetic Variability in Populations*.

[17] Mayr was especially inspired by Conrad H. Waddington's criticism of the synthesis and the centrality of theoretical genetics to the exclusion of other biological fields. See C. H. Waddington, *The Strategy of the Genes* (London: Allen and Unwin, 1957).

fundamental change in thinking about organisms in populational rather than essentialistic or typological terms. Such a view, which stressed the variation between individuals, especially along geographic lines, had come as the result of years of effort on the part of naturalist-systematists dating back to Darwin. Hence the synthesis, which relied so heavily on populational thinking, did not rely exclusively on the contributions of genetics. In a well-known essay "Where Are We?" that he delivered as the inaugural lecture for the 1959 Cold Spring Harbor Symposium entitled "Genetics and Twentieth Century Darwinism" (also celebrating the centennial of Darwin's *Origin*) to an audience that included some of the same mathematical geneticists, Mayr stated pointedly that geneticists had erroneously credited themselves for single-handedly contributing to the populational thinking that made possible the synthetic theory of evolution. For Mayr, such erroneous statements indicated that it was time to "look back without passion or prejudice and determine the respective contributions of experimental genetics, of systematics, of paleontology, of developmental biology, and of other branches of biology, to the synthetic theory of evolution." This same historical spirit of inquiry would help assess contributions of respective groups and plan future work. He wrote: "Such an emphasis on history may be a wholesome counterweight against the exceedingly unhistorical attitude of our current age."[18] Surveying the recent knowledge of the synthetic theory, he concluded that "in spite of the almost universal acceptance of the synthetic theory of evolution, we are still far from fully understanding almost any of the more specific problems of evolution. . . . There is still a vast and wide open frontier."[19]

As Mayr's own historical and philosophical interests grew, so too did the history of biology, genetics, and evolution. By the mid-1960s historical interest in the synthesis was stimulated by the appearance of a series of survey-type histories of genetics written by geneticists L. C. Dunn, A. H. Sturtevant, and Sewall Wright.[20] The occasion of the one hundreth anniversary of the publication of Mendel's "Experiments on Plant Hybrids" led to a special Mendel Centennial Symposium sponsored by the Genetics Society of America in 1965. The proceedings volume, entitled

[18] Ernst Mayr, "Where Are We?" *Cold Spring Harbor Symposia on Quantitative Biology* 24 (1959): 1–14. Quotation on p. 3.
[19] Ibid., p. 13.
[20] L. C. Dunn, *A Short History of Genetics* (New York: McGraw-Hill, 1965); A. H. Sturtevant, *A History of Genetics* (New York: Harper and Row, 1965); Sewall Wright, "The Foundations of Population Genetics," in R. Alexander Brink, ed., *Heritage from Mendel* (Madison: University of Wisconsin Press, 1967), pp. 245–63. These were predated by a suite of classic articles made widely available in 1959: James A. Peters, ed., *Classic Papers in Genetics* (Englewood Cliffs, N.J.: Prentice-Hall, 1959).

Heritage from Mendel, appeared shortly thereafter.[21] A special "Mendel source book," including a translation of his original publication, scientific letters, and articles assessing Mendel and the "rediscovery" of his work, was made available the next year.[22] Later in the decade the history of genetics received a further boost from a younger generation who had scientific training in genetics, but who pursued its history full-time or very nearly so. Robert Olby published the prehistory of Mendelian genetics in his 1966 *Origins of Mendelism*; Garland Allen began his studies of the history of classical genetics culminating with a model for future scientific biographies, *Thomas Hunt Morgan: The Man and His Science*; and Mark Adams brought to the history of genetics a comparative perspective by studying the development of Russian population genetics practically rendered extinct by Lysenko's decree.[23] Others like Elof Axel Carlson traced key scientific concepts in books like *The Gene: A Critical History* or wrote scientific biographies of their mentors like *Genes, Radiation and Society: The Life and Work of H. J. Muller.*[24]

As the mid-1960s witnessed a surge of interest in the history of biology and the history of science, both became recognized academic fields of scholarship in their own right, and the way began to be paved for historical study of the synthesis. In 1968 the first journal devoted exclusively to the history of biology (the *Journal of the History of Biology*) appeared.[25] One reason for beginning the new journal was to attempt to redress an imbalance in the historical literature in favor of the physical sciences.[26] The journal was founded through the efforts of Harvard University-based historians of science who included Ernst Mayr. With the

---

[21] R. Alexander Brink, ed., *Heritage from Mendel* (Madison: University of Wisconsin Press, 1967).

[22] Curt Stern and Eva R. Sherwood, eds., *The Origin of Genetics: A Mendel Source Book* (San Francisco: W. H. Freeman, 1966).

[23] Robert C. Olby, *Origins of Mendelism* (New York: Schocken, 1966); Garland Allen, *Thomas Hunt Morgan: The Man and His Science* (Princeton: Princeton University Press, 1978); Mark B. Adams, "The Founding of Population Genetics: Contributions of the Chetverikov School, 1924–1934," *Journal of the History of Biology* 1 (1968): 23–39; and idem, "Towards a Synthesis: Population Concepts in Russian Evolutionary Thought, 1925–1935," *Journal of the History of Biology* 3 (1970): 107–29. There were few scholarly histories of twentieth-century evolution: Philip G. Fothergill, *Historical Aspects of Organic Evolution* (London: Hollis and Carter, 1952).

[24] Elof Axel Carlson, *Genes, Radiation and Society: The Life and Work of H. J. Muller* (Ithaca: Cornell University Press, 1981). See also idem, *The Gene: A Critical History* (Philadelphia: W. B. Saunders, 1966).

[25] Contributory articles came from Frank N. Egerton, Mark B. Adams, Stephen Jay Gould, Judith P. Swazey, Frederick B. Churchill, Garland E. Allen, and M. D. Grmek. The volume included an essay review by R. C. Lewontin.

[26] Everett Mendelsohn, "Editorial Foreword," *Journal of the History of Biology* 1 (1968): iii–iv.

encouragement of Mayr, then on the board of syndics of Harvard University Press, Everett Mendelsohn began his editorship of the first volume of the new journal.[27]

Many of the early histories of biology upheld what historians of science now frequently deride as the "great man of science" approach, choosing to both recognize and valorize the contributions of scientists to hagiographic excess. As well, philosophical assumptions about the nature of science held to cumulative growth models of scientific knowledge, so that the historiography represented was considered "Whig," another term of derision for many historians of science.[28] Charles Darwin was the most visible historiographic object, and understandably the origins, reception, and exposition of his work received the greater attention from historians of biology. So great was the attention lavished on Darwin that he became associated with a veritable industry of workers called the "Darwin Industry."[29] Of recent biological fields, genetics appeared to draw the most historical attention—as did geneticists—though standard, well-worn historical areas like embryology and physiology also received some attention, especially from historians of the nineteenth-century life sciences. But discussion of the more recent history of evolution and the rise of the synthetic theory of evolution received little historical attention at this time.

Not until 1971, when a compact book entitled *The Origins of Theoretical Population Genetics* made its appearance, did formal historical discussion on the synthesis begin.[30] Written by historian of science William

[27] Judith P. Swazey served as assistant editor.

[28] For examples of the scientific response to the derision given "Whig" history, and for an introduction to a more recent manifestation of the controversy, see Edward Harrison, "Whigs, Prigs, and Historians of Science," *Nature* 329 (1987): 213–14; and Ernst Mayr, "When Is Historiography Whiggish?" *Journal of the History of Ideas* 51 (1990): 301–9. For the original essay on "Whig" history, see H. Butterfield, *The Whig Interpretation of History* (New York: W. W. Norton, 1965, orig. pub. 1931). For further explication and criticism of Butterfield's use of Whiggism, see A. Rupert Hall, "On Whiggism," *History of Science* 21 (1983): 45–59; for an explication of Whig history and presentism or present-centered history, see T. G. Ashplant and Adrian Wilson, "Whig History and Present-Centred History, Part 1," and "Present-Centred History and the Problem of Historical Knowledge, Part 2," *The Historical Journal* 31 (1988): 1–16, 253–74.

[29] Timothy Lenoir, "The Darwin Industry," *Journal of the History of Biology* 20 (1987): 115–30. For a not-so-flattering look at some of the Darwin Industry, see Michael Ruse, "Darwin as Hollywood Epic," *The Quarterly Review of Biology* 61 (1986): 509–15. Without much historical surprise, Alfred Russel Wallace did not receive his due share of attention from historians of evolution.

[30] William B. Provine, *The Origins of Theoretical Population Genetics* (Chicago: University of Chicago Press, 1971). Provine's analysis had been preceded by Mark Adams' work on the contributions of the Russian population geneticists. See Mark Adams, "The Founding of Population Genetics: Contributions of the Chetverikov School 1924–1934," *Journal of the History of Biology* 1 (1968): 23–39.

Provine, the book offered a historical account of many of the critical events that led up to the modern synthesis. A student of Richard C. Lewontin then at the University of Chicago, Provine had studied mathematics and theoretical population genetics extensively before turning his attention to the contributions of theoretical population genetics to evolution in the twentieth century. Provine's subsequent historical interpretations therefore closely followed the geneticists' view that the synthetic theory of evolution resulted from the contributions of mathematical population geneticists Haldane, Fisher, and Wright. Specifically, his historical analysis argued that it was the construction of each of their mathematical models that demonstrated the efficacy of natural selection under a range of different parameters. This made possible the synthesis of genetics and selection theory, which in turn led to the reconciliation between Mendelians and biometricians, who had opposed each other at the turn of the century.[31] At length, Provine examined reasons for these divisions between biologists, which included methodological, personal, intellectual, and scientific animosities.[32] Understanding the removal of the barriers between the participants and their fields of study was the critical key to understanding the origin of the synthetic theory of evolution. While he indicated that some of the more contentious characters died in the process, Provine also pointed out that the construction of the mathematical models was paramount to developments leading to the modern synthesis of evolution in the 1930s. In addition to providing a useful historical road map of the significant events, underscoring the central texts, and introducing one of the more colorful cast of characters in the history of science, Provine's book served to draw attention to the importance of the synthesis in the history of biology. Others began to follow his lead in exploring the more recent history of evolution and redirected their interests accordingly.

Though the book was well received, at least one critic, Ernst Mayr, viewed the interpretation that Provine offered as stressing too heavily the contributions of mathematical population genetics to the detriment of other fields that appeared to play equal, if not greater, roles in the modern synthesis of evolution. Stressing the role of systematics and its contributions to populational thinking, Mayr responded to Provine's interpretation of the more recent history of evolution in an essay review that

---

[31] At the same time, Mendelians abandoned their saltationist views, which had followed from their support of Hugo de Vries's mutation theory. See Provine, *The Origins of Theoretical Population Genetics*, for full explication.

[32] Provine especially focused his historical analysis on the dispute between Mendelians and biometricians at the turn of the century.

coupled Provine's book with Robert Olby's earlier *Origins of Mendelism*. Entitled "The Recent Historiography of Genetics," the essay review appeared in 1973 and applauded efforts by historians to understand biological subjects.[33] But while it served to praise both recent attempts by historians of genetics, especially because they did not represent the genetics establishment, it also reminded readers that numerous sides to the story of recent evolution existed, all of which needed proper formal examination. Mayr's message effectively reinforced his earlier "Where Are We?" lecture of 1959 by stating that the recent history of biology was being overshadowed by the recent histories of genetics. This was especially problematic to proper historical understanding of the synthesis, an event that had relied heavily on not only the work of the mathematical population geneticists, but also on the work of naturalist-systematists and other related biologists. Understanding a major event like the synthesis required consideration of mathematical population genetics as well as the work of naturalists and others who had worked with natural populations of organisms. The problem that had faced many of these naturalists was related to the origin and maintenance of the diversity of life, a problem that had in fact been central to naturalist-systematists.

The response to Provine was extended further in 1974 when Mayr drew together a selected group of original participants, diverse evolutionary biologists, and historians and philosophers of science both to explore the synthesis and to record relevant material for future use by historians.[34] With a clearly stated intent "to tickle the memory of the participants and to give them an opportunity to elaborate on my comments or correct them if my recollections or interpretations are faulty," Mayr organized two successive special workshops held 23–25 May and 11–12 October in 1974 at the American Academy of Arts and Sciences.[35] Detailed transcripts of conference discussions were made, and many participants filled

---

[33] See E. Mayr, "The Recent Historiography of Genetics," *Journal of the History of Biology* 6 (1973): 125–54.

[34] These had been the aims clearly stated in the proposal submitted to the National Science Foundation: "The proposed workshop is a preliminary effort to assemble materials relevant to the modern synthesis in evolutionary biology and to organize data in a way that will make it most accessible to historians of twentieth century science." Quotation on p. 4 of an NSF Proposal written by Ernst Mayr and John Voss (Institutional Executive Officer, American Academy of Arts and Sciences); submission of proposal dated 4 January 1974. In Library of the American Philosophical Society, Ernst Mayr Papers, Box A–P, "Evolutionary Synthesis Conference."

[35] This had taken place through "The Committee on the Recent History of Science and Technology" at the American Academy of Arts and Sciences. The historical documentation is at the Library of the American Philosophical Society, catalogued with the "Ernst Mayr Papers" under the "Evolutionary Synthesis Conference."

out special historical questionnaires for future reference. Conference materials that were generated were duplicated and most were subsequently deposited in the largest American repository for historical materials on genetics and evolution, the archival collections of the American Philosophical Society, the very society that had provided funds for the start-up of the new SSE's journal, *Evolution*. With the assistance of Provine, Mayr edited an ambitious volume entitled *The Evolutionary Synthesis: Perspectives on the Unification of Biology*, published by Harvard University Press.[36] Based on many of the materials that they had solicited, the volume included original contributions by the various participants, commentaries by Mayr and Provine, and historical reflections and notes by some of the original participants. A master list of workshop participants, the contributors to the volume, and recent addresses was included for future reference.[37] The workshop discussions, the compilation of historical materials, and the completed volume were so successfully orchestrated that as interest in the synthesis grew, Mayr's appellation of "the evolutionary synthesis" began to gain historical currency, designating the historical event in question.

The volume also served to highlight the differences of opinion between Mayr and Provine. For Mayr, the work of naturalist-systematists and the rise of populational thinking continued to play a prominent historical role in the evolutionary synthesis. So, too, Mayr stressed the articulation of the biological species concept and what this entailed in understanding geographic speciation and his and Dobzhansky's role in articulating it first in 1937 (for Dobzhansky) and then 1942 (for Mayr). Provine, on the other hand, continued to stress the importance of mathematical population geneticists and to focus on how barriers to the synthesis were effectively removed between fields and participants,[38] though he also indicated in his epilogue and elsewhere that the confer-

[36] Ernst Mayr and William B. Provine, eds., *The Evolutionary Synthesis: Perspectives on the Unification of Biology* (Cambridge, Mass.: Harvard University Press, 1980). This remains the most comprehensive work on the evolutionary synthesis.

[37] The first workshop included Ernst Mayr (as chairman), Mark B. Adams, Ernest Boesiger, Peter Buck, Richard W. Burkhardt, Jr., Hampton L. Carson, William Coleman, C. D. Darlington, Theodosius Dobzhansky, E. B. Ford, Bentley Glass, Stephen Jay Gould, Gerald Holton, David L. Hull, E. David Kohn, I. Michael Lerner, Richard C. Lewontin, Camille Limoges, William B. Provine, Dudley Shapere, Otto Solbrig, G. Ledyard Stebbins, Frank Sulloway, Robert L. Trivers, and Alexander Weinstein. The second workshop included Ernst Mayr (as chairman), Mark B. Adams, Garland Allen, Frederick Churchill, William Coleman, Irven De Vore, Michael T. Ghiselin, Stephen Jay Gould, John C. Greene, Wayne Gruner, Viktor Hamburger, E. David Kohn, Richard C. Lewontin, Camille Limoges, Everett C. Olson, Ronald Overmann, William B. Provine, Bobb Schaeffer, Dudley Shapere, Frank Sulloway, Daniel Todes, Robert L. Trivers, and Alexander Weinstein.

[38] See also William B. Provine, "The Role of Mathematical Population Geneticists in the

ence had generated a greater range of complex issues for further consideration.

Although these two workshops convinced the participants and subsequent workers of the importance of the evolutionary synthesis and of the need for further study, it also brought into relief the disparity of opinions that existed over what exactly constituted the modern synthesis of evolution.[39] Original participants, many of whom were still alive at the time, took especial issue with historical interpretations that were not in agreement with their firsthand experience of events. Interestingly enough, these original participants felt to varying degrees that their own respective fields had not received proper historical recognition. Very often they indicated that their own contributions had been unappreciated. Among the original participants voicing their complaints (in addition to Mayr) were G. G. Simpson, who did not personally attend the conference because of ill health, but also because he felt that it was not the proper forum in which to assess his own contributions (he subsequently did this in his autobiography *Concession to the Improbable* in 1978),[40] and G. Ledyard Stebbins, who reminded his colleagues of the animal-centeredness of the synthesis to the exclusion of the contributions of botanists.[41] Sewall Wright had proper reason to feel excluded, as he was not

---

Evolutionary Synthesis of the 1930s and 1940s," *Studies in History of Biology* 2 (1978): 167–92.

[39] See Mayr and Provine, eds., *The Evolutionary Synthesis*.

[40] G. G. Simpson, *Concession to the Improbable: An Unconventional Autobiography* (New Haven: Yale University Press, 1978). The extensive correspondence between Simpson and Mayr over the evolutionary synthesis conference is included with the conference correspondence. From this correspondence it is clear that Mayr made numerous attempts to include Simpson, though Simpson repeatedly came up with reasons for not attending. Simpson's unwillingness to attend did not deter him from sharply criticizing the discussions in the transcripts sent to him following the conference by Mayr. He wrote: "It is both interesting and disheartening to read what some colleagues think of my thoughts and work in what I suppose is a fair report of frank and informal discussion. It is disheartening especially because on the whole they don't seem to have understod what in fact I thought and wrote, and indeed do not seem to have made an honest effort to do so before sounding off." He continued: "Perhaps, however, if my views had been better understood and represented they would be even less appreciated" (letter from G. G. Simpson to Ernst Mayr, dated 26 July 1975). Anne Roe Simpson also responded sharply to Mayr on the same day: "I am sorry to say that it seems to me to have been rather less than worth the time and effort, not to say the money that went into that particular meeting." She continued: "I comment from the standpoint of one who has made the study of careers a major interest, and I feel that to have approached the problem of the development of a scientific point of view in such a way has been to do a great disservice to an accurate history of the development" (letter from Anne Roe Simpson to Ernst Mayr, dated 26 July 1975). In Library of the American Philosophical Society, Ernst Mayr Papers, Box Q–Z, "Conference on the Evolutionary Synthesis," Simpson folder.

[41] G. Ledyard Stebbins, "Botany and the Synthetic Theory of Evolution," in E. Mayr and W. B. Provine, eds., *The Evolutionary Synthesis: Perspectives on the Unification of Biology* (Cambridge, Mass.: Harvard University Press, 1980), pp. 139–52.

even invited to attend;[42] and responding to the historical call, Conrad H. Waddington, one of the earliest critics of the synthesis, eventually guaranteed that his own contributions were appreciated by writing *Evolution of an Evolutionist* in 1975.[43] Recounting the conference and analyzing the feelings of the original participants on their contributions or the contributions of their respective fields later in 1988, Provine noted a marked symmetry of response: "The symmetry here is ever so reminiscent of an academic department of top-notch scholars, each of whom predictably believes that his or her work is not properly appreciated by the chairman and the rest of the department, no matter what rewards, accolades, and support are provided."[44]

Evolutionary biologists who were preoccupied by current developments in evolutionary theory read many of the then current debates back into the history of the synthetic theory. Among the most prominent of these was Stephen Jay Gould, who added yet another interpretive twist by arguing that the synthesis had become *too* firmly established, especially because of its overly strong reliance on the mechanism of natural selection. Arguing that in its later phases the synthesis led to extreme pan-selectionism, Gould stated that the evolutionary synthesis had "hardened" to such an extent that it hindered proper consideration of the complexity of evolutionary processes that included more prominent roles for nonadaptive mechanisms of evolution and other considerations such as the role of developmental constraints.[45] Other attendees, like the phi-

---

[42] According to Mayr, Wright was not invited to attend because: (a) he was too senior, (b) he had a tendency to be verbose in public settings, and (c) there were too many geneticists (some twenty-two) who had already been invited. See Ernst Mayr's reasons for his dispute with Wright in Ernst Mayr, "Controversies in Retrospect," in Douglas Futuyma and Janis Antonovics, eds., *Oxford Surveys in Evolutionary Biology* 8 (1992): 1–34. Reasons for the differences between Mayr and Wright, including Wright's sense that his emphasis on random genetic drift was given too little historical credit, are also discussed in chap. 12 of William B. Provine, *Sewall Wright and Evolutionary Biology* (Chicago: University of Chicago Press, 1986). According to Provine, Wright had written a "strongly worded" review of Mayr's 1959 "Where Are We?" address: see Sewall Wright, "Genetics and Twentieth Century Darwinism: A Review and Discussion," *American Journal of Human Genetics* 12 (1960): 365–72. Julian Huxley and Bernhard Rensch could not attend because of illness and advanced age. Rensch sent his contribution for inclusion in the edited volume on Neo-Darwinism in Germany.

[43] Conrad H. Waddington, *Evolution of an Evolutionist* (Ithaca: Cornell University Press, 1975).

[44] William B. Provine, "Progress in Evolution and Meaning in Life," in Matthew H. Nitecki, ed., *Evolutionary Progress* (Chicago: University of Chicago Press, 1988), pp. 49–74. Quotation on p. 55.

[45] Stephen Jay Gould, "G. G. Simpson, Paleontology, and the Modern Synthesis," in E. Mayr and W. B. Provine, eds., *The Evolutionary Synthesis: Perspectives on the Unification of Biology* (Cambridge, Mass.: Harvard University Press, 1980), pp. 153–72. See also Stephen J. Gould, "The Hardening of the Modern Synthesis," in Marjorie Grene, ed., *Dimensions of Darwinism* (Cambridge: Cambridge University Press, 1983), pp. 71–93.

losophers and historians of science, additionally grappled with the inadequacy of any existing philosophical framework or a general theory of science that could shed light on how to best view the synthesis in the wider picture of the history and philosophy of science. After consideration of what could count as a proper synthesis or unification within the sciences, Dudley Shapere wrote: "All this is to say that the usual ways of thinking about scientific change and innovation do not quite capture the entirety of achievement of the synthetic theory of evolution, which in turn suggests that there is something special to be learned about the nature of scientific change and the scientific enterprise from this case that cannot be learned from most other cases."[46]

Further complicating historical understanding were the national contexts for evolutionary activity represented at the workshop. Nations as different as the Soviet Union, Germany, France, England, and the United States held to their respective contributions and historical interpretations. Historical exploration of the national contexts of activity made it apparent that the synthesis appeared to take radically different historical directions in different national contexts. Transcripts of the conference indicate that there appeared to be as many issues raised, and as many interpretative twists and turns, as there were conference participants to represent points of view. But from the epilogue to the volume that appeared in 1980, Provine indicated that despite all the differences of opinion brought into relief by the conference, "All participants, whether scientists or historians, young or old, agreed that a consensus concerning the mechanism of evolution appeared among biologists during the 1920–1950 period."[47] Participants seemed to agree at least on this point.

Ironically, the same year that saw the publication of this volume in honor of the evolutionary synthesis also witnessed the first of what became explosive developments in evolutionary theory. Thus, at the same time that some were assessing the historical contributions of the evolutionary synthesis, others began to point to its inadequacies. Debates among evolutionists over the mechanism of evolution and other integral portions of evolutionary theory brought into relief not only unsolved problems in the synthetic theory, but began to question its validity as a theory and its veracity as a proper "synthesis" between biological disciplines. As evolutionary biologists became divided over the most funda-

[46] Dudley Shapere, "The Meaning of the Evolutionary Synthesis," in E. Mayr and W. B. Provine, eds., *The Evolutionary Synthesis: Perspectives on the Unification of Biology* (Cambridge, Mass.: Harvard University Press, 1988), pp. 388–98.

[47] William B. Provine, "Epilogue," in Mayr and Provine, eds., *The Evolutionary Synthesis*, p. 399.

mental tenets of the synthetic theory, the synthetic theory itself appeared to crumble.

The most public of these debates drew the attention of evolutionary biologists during a conference at the Chicago Field Museum of Natural History in 1980 with the conference title of "Macroevolution." Though the "punctuated equilibrium" theory of evolution had been introduced earlier in 1972 and then in 1977 to smaller circles of paleontologists,[48] the conference on macroevolution focused attention on the controversial theory (and its proposed amendments), drawing the increasing interest of evolutionary biologists outside the paleontological community. Because, too, it was lucidly reported in widely read scientific journals like *Science* and *Nature* (with suitably sensationalist titles like "Evolutionary Theory under Fire")[49] and even in popular magazines like *Newsweek* (whose running title was "Enigmas of Evolution"),[50] it also drew the attention of a wider audience that included both scientists and lay readers. As the next generation of postsynthesis evolutionary biologists emerged at this time, evolutionary theory was called into question as it was closely reexamined. As the synthesis came under close scrutiny and criticism, and as the rhetoric of debate accelerated, the coherence of the synthetic theory increasingly appeared to diminish. A rash of what can be described as "antisynthesis" literature began to appear in the 1980s as some detractors held that the synthetic theory of evolution was appearing to unravel in the light of new developments in evolution. Among the points made by commentators was that the synthesis was incomplete, misdirected, or just plain wrong. Others responded to what they viewed as criticisms, threats, and/or challenges by defending the status of the synthetic theory; others still tried to develop conciliatory positions. A quick survey of just some of the titles of reviews, commentaries, articles, and books that appeared during this time captures the increasing unease,

---

[48] Niles Eldredge and S. J. Gould, "Punctuated Equilibria: An Alternative to Phyletic Gradualism," in T. J. M. Schopf, ed., *Models in Paleobiology* (San Francisco: Freeman Cooper, 1972), pp. 82–115; S. J. Gould and Niles Eldredge, "Punctuated Equilibria: The Tempo and Mode of Evolution Reconsidered," *Paleobiology* 3 (1977): 115–51. For the most recent collection of essays on "punc eq" debates, see Albert Somit and Steven A. Peterson, eds., *The Dynamics of Evolution: The Punctuated Equilibrium Debate in the Natural and Social Sciences* (Ithaca: Cornell University Press, 1992). For a historical analysis, see Neal Andrew Doran, "Punctuating Evolution: The Metaphysical Foundations of the Punctuated Equilibrium Controversy" (Master's thesis, University of Florida, 1994).

[49] Roger Lewin, "Evolutionary Theory under Fire," *Science* 210 (1980): 883–87. The subtitle gave the conference historic dimensions: "An historic conference in Chicago challenges the four-decade long dominance of the Modern Synthesis." This was nearly a five-page "Research News" column in *Science*.

[50] Jerry Adler and John Carey, "Enigmas of Evolution," *Newsweek* 99 (1982): 44–49.

dissatisfaction, defensiveness, and general confusion that surrounded discussions of the synthesis in the 1980s: "Is a New and General Theory of Evolution Emerging?"; "Is a New Evolutionary Synthesis Necessary?"; "The Evolutionary Synthesis Is Only Partially Wright" (the response was "But Not Wright Enough"); "Beyond Darwinism? The Challenge of Macroevolution to the Synthetic Theory of Evolution"; "Beyond Neo-Darwinism: An Introduction to the New Evolutionary Paradigm"; "Evolutionary Theory: The Unfinished Synthesis"; "The Unfinished Synthesis: Biological Hierarchies and Modern Evolutionary Thought"; "Evolution at a Crossroads: The New Biology and Philosophy of Science"; "Challenges to the Evolutionary Synthesis"; "The Evolutionary Dys-Synthesis: Which Bottles for Which Wine?"; "More Darwinian Detractors"; "The Synthetic Theory Strikes Back"; "The Triumph of the Evolutionary Synthesis"; and "Darwinism Stays Unpunctured."[51]

Controversies in evolution had gone far beyond just the paleontological reformation of evolution. The application of nonequilibrium thermodynamics and information theory to evolution precipitated a round of debates, some of which were nearly incomprehensible to members of the evolution community.[52] Systematists, fired by Willi Hennigs' manifesto

[51] S. J. Gould, "Is a New and General Theory of Evolution Emerging?" *Paleobiology* 6 (1980): 119–30; G. Ledyard Stebbins and F. J. Ayala, "Is a New Evolutionary Synthesis Necessary?" *Science* 213 (1981): 967–71; S. Orzack, "The Evolutionary Synthesis Is Only Partly Wright," *Paleobiology* 7 (1981): 128–31; S. J. Gould, "But Not Wright Enough: Reply to Orzack," *Paleobiology* 7 (1981): 131–34; F. J. Ayala, "Beyond Darwinism? The Challenge of Macroevolution to the Synthetic Theory of Evolution," *Philosophy of Science Association* 2 (1982): 275–91; Stephen J. Gould, "Darwinism and the Expansion of Evolutionary Theory," *Science* 216 (1982): 380–87; Mae-Wan Ho and P. T. Saunders, *Beyond Neo-Darwinism: An Introduction to the New Evolutionary Paradigm* (New York: Academic Press, 1984); Robert G. B. Reid, *Evolutionary Theory: The Unfinished Synthesis* (Ithaca: Cornell University Press, 1985); Niles Eldredge, *The Unfinished Synthesis: Biological Hierarchies and Modern Evolutionary Thought* (Oxford: Oxford University Press, 1985); D. J. Depew and Bruce. H. Weber, eds., *Evolution at a Crossroads: The New Biology and Philosophy of Science* (Cambridge: MIT Press, 1985); Richard Burian, "Challenges to the Evolutionary Synthesis," *Evolutionary Biology* 23 (1988): 247–69; J. Antonovics, "The Evolutionary Dys-Synthesis: Which Bottles for Which Wine?" *American Naturalist* 129 (1987): 321–31; Mark Ridley, "More Darwinian Detractors," *Nature* 318 (1985): 124–26; V. Grant, "The Synthetic Theory Strikes Back," *Biologische Zentralblatt* 102 (1983): 149–58; Ernst Mayr, "The Triumph of Evolutionary Synthesis," *Times Literary Supplement*, 2 November 1984, pp. 1261–62; John Maynard Smith, "Darwinism Stays Unpunctured," *Nature* 330 (1987): 516. See also Jeffrey W. Pollard, ed., *Evolutionary Theory: Paths into the Future* (New York: Wiley, 1984).

[52] E. O. Wiley and D. R. Brooks, "Victims of History—a Nonequilibrium Approach to Evolution," *Systematic Zoology* 31 (1982): 1–24; Daniel R. Brooks and E. O. Wiley, "Evolution as Entropic Phenomenon," in Jeffrey W. Pollard, ed., *Evolutionary Theory: Paths into the Future* (New York: Wiley, 1984), pp. 141–71; D. R. Brooks amd E. O. Wiley, *Evolution as Entropy: Toward a Unified Theory of Biology* (Chicago: University of Chicago Press, 1986); and D. J. Depew and B. H. Weber, "Consequences of Nonequilibrium Thermodynamics for the Darwinian tradition," in B. H. Weber et al., eds., *Entropy, Information and*

Phylogenetic Systematics (1966), waged an internecine war against rival taxonomic methodologies under the crusading banner of "cladistics."[53] Mayr was quick to defend evolutionary classification in 1981 to readers of *Science*.[54] Even discussions of Lamarckism were revived under the title of "somatic selection" in the early 1980s.[55] Differences of opinion—personal, political, and scientific—over evolutionary theory became even more conflated as they made their way to the British public when biologists objected over the new display of dinosaurs in the British Museum. Blurring cladistics with punctuated equilibrium, Marxism, the politicization of knowledge, and the rising creationist threat, the debates over evolutionary theory made their way to *Science* in 1981 under the title "Dinosaur Battle Erupts in British Museum: Anti-cladist Sees Reds under Fossil Beds in Alliance with Creationists to Subvert Establishment."[56] Finally, controversies surrounding the synthetic theory of evolution were so numerous that by the early 1980s a book-length "guide" to the controversies appeared to help confused observers. As the title, *Darwinism Defended*, suggested, the goal ultimately was to reassure readers that the Darwinian theory was still valid despite all the attacks.[57]

From yet another direction, the synthetic theory began to receive one of its major successful amendments as techniques such as polyacrylamide gel electrophoresis and other insights gleaned from molecular biology made their way into the study of evolutionary processes.[58] Beginning in

---

Evolution: Perspectives on Physical and Biological Evolution (Cambridge: MIT Press, 1988). For a scathing critique, see Harold Morowitz, "Entropy and Nonsense: A Review of Daniel R. Brooks and E. O. Wiley, *Evolution as Entropy*," *Biology and Philosophy* 1 (1986): 473–76. See also the response: E. O. Wiley and Daniel R. Brooks, "A Response to Professor Morowitz," *Biology and Philosophy* 2 (1987): 369–74.

[53] Willi Hennig, *Phylogenetic Systematics*, trans. D. Dwight Davis and Rainer Zangerl (Urbana: University of Illinois Press, 1966). By far the most comprehensive account of these debates is David L. Hull, *Science as a Process: An Evolutionary Account of the Social and Conceptual Development of Science* (Chicago: University of Chicago Press, 1988).

[54] Ernst Mayr, "Biological Classification: Toward a Synthesis of Opposing Methodologies," *Science* 214 (1981): 510–16.

[55] E. J. Steele, *Somatic Selection and Adaptive Evolution: On the Inheritance of Acquired Characters*, 2d ed. (Chicago: University of Chicago Press, 1981; 1st ed., 1979). The scientific press reported the revival of Lamarckism with articles like Roger Lewin, "Lamarck Will Not Lie Down," *Science* 213 (1981): 316–21.

[56] Nicholas Wade, "Dinosaur Battle Erupts in British Museum," *Science* 211 (1981): 35–36.

[57] Michael Ruse, *Darwinism Defended: A Guide to the Evolution Controversies* (Reading, Mass.: Addison-Wesley, 1982). See also D. R. Brooks, "What's Going on in Evolution: A Brief Guide to Some New Ideas in Evolutionary Theory," *Canadian Journal of Zoology* 61 (1983): 2637–45.

[58] The first such use was by Jack L. Hubby, "Protein Differences in *Drosophila*. I. *Drosophila melanogaster*," *Genetics* 48 (1963): 871–79. The papers introducing the technique to measure variation in natural populations were J. L. Hubby and R. C. Lewontin, "A Molecular Approach to the Study of Genic Heterozygosity in Natural Populations. I. The

1968 with the work of Japanese geneticist Motoo Kimura and later with the collaborative work of Jack L. King and Thomas Jukes, what was later termed the neutral theory of molecular evolution was introduced.[59] Architects of the evolutionary synthesis, who had been watching developments in molecular biology with some wariness,[60] were quick to react to the appearance of claims of a "Non-Darwinian Evolution" that did not appear to give natural selection preeminence. By 1983 Kimura claimed in *The Neutral Theory of Molecular Evolution*[61] that the synthetic theory had become so well entrenched (and "overdeveloped") as to represent the "orthodox view." In Kimura's view of the history of twentieth-century evolution, advocates of the synthesis had criticized his neutral theory of molecular evolution to such an extent that they had actually hindered understanding of evolution at the molecular level.

The attention that evolutionary theory received at this time increased further still with the appearance of yet another synthesis of evolution leading to the new field of sociobiology. In a string of books written first by skilled writers like Edward O. Wilson and then Richard Dawkins, sociobiology and biologically informed behavioral study rose in popularity.[62] Richard Dawkins pushed deterministic arguments to their ex-

Number of Alleles at Different Loci in *Drosophila pseudoobscura*," *Genetics* 54 (1966): 577–94; and "II. Amount of Variation and Degree of Heterozygosity in Natural Populations of *Drosophila pseudoobscura*," *Genetics* 54 (1966): 595–609.

[59] Motoo Kimura, "Evolutionary Rate at the Molecular Level," *Nature* 217 (1968): 624–26; Jack L. King and Thomas Jukes, "Non-Darwinian Evolution," *Science* 164 (1969): 788–98. Historical discussion of the neutral theory is included in R. C. Lewontin, *The Genetic Basis of Evolutionary Change* (New York: Columbia University Press, 1974); for more recent historical analysis, see William B. Provine, "The Neutral Theory of Molecular Evolution in Historical Perspective," in Naoyuki Takahata and James Crow, eds., *Population Biology of Genes and Molecules* (Tokyo: Baifukan Press, 1990), pp. 17–31; and Michael R. Dietrich, "The Origins of the Neutral Theory of Molecular Evolution," *Journal of the History of Biology* 27 (1994): 21–59. See also the introductory essays by Naoyuki Takahata in Motoo Kimura, *Population Genetics, Molecular Evolution, and the Neutral Theory: Selected Papers*, ed. and with introductory essays by Naoyuki Takahata, foreword by James F. Crow (Chicago: University of Chicago Press, 1994).

[60] G. G. Simpson, "Organisms and Molecules in Evolution," *Science* 146 (1964): 1535–38.

[61] Motoo Kimura, *The Neutral Theory of Molecular Evolution* (Cambridge: Cambridge University Press, 1983). Chapter 2 was entitled "Overdevelopment of the Synthetic Theory and the Proposal of the Neutral Theory."

[62] Edward O. Wilson, *Sociobiology: The New Synthesis* (Cambridge, Mass.: Belknap, 1975); idem, *On Human Nature* (Cambridge, Mass.: Harvard University Press, 1978); Richard Dawkins, *The Selfish Gene* (Oxford: Oxford University Press, 1976); idem, *The Extended Phenotype: The Gene as the Unit of Selection* (Oxford: Oxford University Press, 1982); idem, *The Blind Watchmaker* (New York: W. W. Norton, 1986); David Barash, *The Whisperings Within* (Harmondsworth: Penguin, 1979); idem, *The Hare and the Tortoise: Culture, Biology and Human Nature* (New York: Viking, 1986); Robert A. Wallace, *The Genesis Factor* (New York: William Morrow, 1979); Melvin Konner, *The Tangled Wing: Biological Constraints on the Human Spirit* (New York: Holt, Rinehart, and Winston, 1982).

treme yet logical end by arguing for the concept of "memes"—units of cultural inheritance on which selection operated. As sociobiology was permuted by both its practitioners and theorists, and as it appeared to embrace more and more of a reductionist program that sought to understand behavior and society in terms of genetic determinants, sociobiology generated extensive controversy that spread far beyond the smaller circles of evolutionary biologists.[63] Equally skilled writers like Stephen Jay Gould and Richard C. Lewontin argued forcefully against what they viewed as the simplistic reductionism and sociopolitical dangers inherent in the sociobiological program. As a result, the growing interest in the validity of the synthetic theory, questions on the extent to which selectionism operated, and whether or not sociobiology stemmed logically from the synthetic theory (as was argued by Wilson) served to draw further attention to the evolutionary synthesis.[64] At stake now was no longer the status of just an arguably "abstract" scientific theory, but a sociopolitical program of action.[65] Nor was this limited to the internal debates of biologists: Wilson's eloquence reached vast audiences, eventually earning him two Pulitzer Prizes.[66] Stephen Jay Gould's natural history essays, historical works, and other writings garnered an equally vast popular readership, all of whom were introduced to the current debates in evolution.

Adding to the sociopolitical burden of evolutionary biology were the moral, ethical, and judicial concerns brought on by the fundamentalist challenge to the teaching of evolution in American high schools. In the

[63] Wilson explicitly denied that he was a strict genetic determinist. See his discussion in *On Human Nature*.

[64] Stephen J. Gould and R. C. Lewontin, "The Spandrels of San Marco and the Panglossian Paradigm: A Critique of the Adaptationist Programme," *Proceedings of the Royal Society of London* B 205 (1979): 581–98; Stephen J. Gould, "Sociobiology: The Art of Story-Telling," *New Scientist* 80 (1978): 530–33; idem, "Sociobiology and the Theory of Natural Selection," in G. W. Barlow and J. Silverberg, eds., *Sociobiology: Beyond Nature/Nurture?* (Boulder, Colo.: Westview, 1980), pp. 257–69; R. C. Lewontin, "Sociobiology: A Caricature of Darwinism," *Philosophy of Science Association* 2 (1976): 22–31; idem, "Biological Determinism as a Social Weapon," in The Ann Arbor Science for the People Editorial Collective, ed., *Biology as a Social Weapon* (Minneapolis: Burgess, 1977), pp. 6–18; and idem, "Sociobiology as an Adaptationist Program," *Behavioral Science* 24 (1979): 5–14. See also R. C. Lewontin, S. Rose, and L. Kamin, *Not in Our Genes: Biology, Ideology, and Human Nature* (New York: Pantheon, 1984).

[65] Even the long-held traditional view of the political neutrality (and hence the transcendence) of science was eventually abandoned as biology became overtly politicized. Among the abundant literature on the political dimensions of biology, see Richard Levins and Richard C. Lewontin, *The Dialectical Biologist* (Cambridge, Mass.: Harvard University Press, 1985). The dedication page read: "To Frederick Engels, who got it wrong a lot of the time but who got it right where it counted."

[66] Wilson received a Pulitzer Prize for *The Ants*, which was published in 1990, and for *On Human Nature*, which was published in 1978.

late 1970s and early 1980s fundamentalist groups devised a new strategy to win their war on the teaching of evolution in American public schools. Rather than simply arguing against evolution, they argued additionally for their properly "scientific" version of creationism. In the 1980s fundamentalists under the banner of "scientific creationism" launched a series of public attacks against evolutionary theory, arguing for "equal time" for creationist accounts. In these disputes, scientific creationists did not hesitate to cash out on some of the most vituperative debates on the synthetic theory to demonstrate the inadequate methodological and indeed scientific grounding of contemporary evolution.[67] Despite evolutionists' repeated statements to the effect that evolution itself was a fact, and that only consideration of possible mechanisms was generating debate, evolutionists challenging portions of the synthetic theory also unwittingly armed the creationist arsenal. As the teaching of evolution was contested in states like California, Arkansas, and Louisiana, it became a public courtroom drama; and as noted evolutionists testified for the validity of evolutionary theory and the proper methodology of science, evolution became an even greater subject for newsworthy public debate.[68] By 1980 Ronald Reagan, then merely a presidential hopeful, received applause and support from fundamentalists and religious others of the New Right when he publicly undermined the status of evolution. To an audience numbering approximately ten thousand fundamentalists in Dallas, Texas, he responded to questions about his belief in evolution by stating: "Well, it is a theory, it is a scientific theory only, and it has in recent years been challenged in the world of science and is not yet believed in the scientific community to be as infallible as it once was believed. But if it was going to be taught in the schools, then I think that also the biblical theory of creation, which is not a theory but the biblical story of creation, should also be taught."[69]

[67] Creationists were especially drawn to the sensationalist titles generated by journalists covering the debates. A very recent letter to the editor of *Science*, entitled "Darwin in the Headlines," written by the director of the Biological Sciences Curriculum Study, pointed out the dangers of such journalistic use of titles and its effects on fueling the evolution–creation controversies: Joseph D. McInerney, "Letter to the Editor, Darwin in the Headlines," 5 May 1995, *Science* 268 (1995): 624.

[68] For the most recent historical analysis, see George E. Webb, *The Evolution Controversy in America* (Lexington: University Press of Kentucky, 1994). See also Ronald L. Numbers, *The Creationists: The Evolution of Scientific Creationism* (Berkeley: University of California Press, 1992).

[69] As cited in Webb, *The Evolution Controversy in America*, p. 217. For more details, see the discussion on pp. 216–17. See also Jeffrey L. Brudney and Gary W. Copeland, "Evangelicals as a Political Force: Reagan and the 1980 Religious Vote," *Social Science Quarterly* 65 (1984): 1072–79.

As if evolution were not receiving enough attention, developments in anthropology, the science that directly addressed human evolution, altered its understanding of the larger picture of hominid evolution by the early 1980s. The discovery of *Australopithecus afarensis*, affectionately nicknamed "Lucy," by Donald Johanson and his colleagues in the 1970s was nothing short of a spectacular find.[70] What should have been cause for public celebration and fruitful discussion on the origins of humans became the occasion for scientific rivalry as Johanson and Richard E. Leakey, heir to an anthropological legacy, crossed swords publicly.[71] In front of their host, Walter Cronkite, and a television audience of millions waiting to hear of the new developments, Leakey revealed what he really thought of Johanson's model of human evolution by drawing a large black "X" over the poster figure representing Johanson's proposed lineage for hominid evolution. With this one quick gesture of dismissal, debates within the evolutionary community were reduced to an embarrassing public spectacle.[72]

Understanding of the synthesis itself became even more complicated as a new generation of philosophers turned to biology at this time.[73] The publication in 1959 of the notorious critique of evolution by Marjorie Grene had drawn attention to the suitability—and vulnerability—of evolutionary theory to philosophical critique.[74] As Grene turned to full-time study of evolution, the dialogue with architects like G. Ledyard Stebbins convinced her of both the unusualness and usefulness of this new science for philosophical exploration. In so doing she began to generate a corpus of literature (and not a few followers) on the new philosophy of science that focused on biology. At the same time, Ernst Mayr began to pick up on the heightened interest in the philosophy of science, and called for a philosophical exploration of evolution and biology, sciences that held to

[70] See Donald C. Johanson and Maitland A. Edey, *Lucy: The Beginnings of Humankind* (New York: Simon and Schuster, 1981).

[71] Richard Leakey consciously strove to transmit developments to a public audience. In the late 1970s and early 1980s he joined with writer Roger Lewin to write semipopular books on evolution. See Richard E. Leakey and Roger Lewin, *Origins* (New York: E. P. Dutton, 1977); and idem, *People of the Lake: Mankind and Its Beginnings* (Garden City, N.Y.: Anchor Press/Doubleday, 1978).

[72] Johanson discusses the nature of the dispute with Leakey on pp. 298–306 in *Lucy*.

[73] For more information on the background of many philosophers of biology, see the recent volume organized and moderated by Werner Callebaut, *Taking the Naturalistic Turn or How Real Philosophy of Science Is Done* (Chicago: University of Chicago Press, 1993). This volume includes interviews and personalia on established figures in the philosophy of biology. For a recent survey of the field, see Kim Sterelny, "Understanding Life: Recent Work in Philosophy of Biology," *British Journal for the Philosophy of Science* 46 (1995): 155–83.

[74] Marjorie Grene, "Two Evolutionary Theories," *British Journal for the Philosophy of Science* 9 (1959): 110–27, 185–93.

logical principles different from the philosophical exemplar science of physics. As they responded to Mayr's call for a new philosophy of science, hopeful philosophers began to look to biology to derive new laws and procedures in science, as well as to help out with some of the more complex yet important recent problems in science. Following Mayr's lead on the subject, and because the belief was widely held that evolutionary theory effectively "unified" the biological sciences, philosophers especially focused on evolutionary theory for their philosophical inquiries. No strangers to controversy, philosophers were especially attracted to the debates on the validity of evolutionary theory as a scientific theory.[75] In search of training in evolutionary theory (for them a new area of inquiry), an entire generation of philosophers of biology turned for their mentorship to the charismatic young heir to synthesis, Richard C. Lewontin. As the numbers of philosophers coming through "Lewontin's lab" increased, so too did the interest in philosophical examination of evolution—especially on the current debates involving Lewontin. As they acquainted themselves with logical convolutions and the controversies in evolutionary theory, some philosophers developed a sophisticated rationale for their growing interest in evolution. As Michael Ruse explained in a crisp synopsis of this "hot topic" in *Philosophy of Biology Today*: "Philosophical inquiry begins when, and to a certain extent only when, scientists fall out. Just as the paleontologists Gould and Eldredge certainly like their theory, in part, because it makes paleontology vital in the understanding of mechanisms, so philosophers like theories which raise philosophical issues."[76]

As the numbers of philosophers working on biology increased, and as they examined philosophy of evolution and biology, philosophers began to generate copious quantities of literature to describe the "structure" of evolutionary theory as well as the "structure" of biology.[77] Evolutionary

[75] Building on his earlier critique of historical knowledge, *The Poverty of Historicism*, Karl Popper turned his critical lens to evolutionary theory and questioned its scientific validity in the mid-1970s. Karl Popper, "Darwinism as Metaphysical Research Program," in *The Philosophy of Karl Popper*, ed. P. Schilpp (La Salle, Ill: Open Court, 1974). See also Karl Popper, *The Poverty of Historicism*, 3d ed. (New York: Harper Torchbooks, 1966; orig. pub. 1944). For a discussion of Popper and evolutionary theory, see Robert J. Richards, "The Structure of Narrative Explanation in History and Biology," in Matthew H. Nitecki and Doris V. Nitecki, eds., *History and Evolution* (Albany: State University of New York Press, 1992), pp. 19–53.

[76] Michael Ruse, *Philosophy of Biology Today* (Albany: State University of New York Press, 1988), p. 38.

[77] Michael Ruse, "Is the Theory of Evolution Different? I. The Central Core of the Theory. II. The Structure of the Entire Theory," *Scientia* 106 (1971): 765–83, 1069–93; R. N. Brandon, "A Structural Description of Evolutionary Theory," in P. Asquith and R. Giere, eds., *Philosophy of Science Association* 2 (1981): 427–39. See also Arthur L. Cap-

theory, they quickly discovered, defied any model of a scientific theory derived from examples in the physical sciences. As philosophers created special terminology to describe the peculiar scientific theory, evolution was described as a "supratheoretical framework," a "hypertheory," or a "metatheory."[78] While some disputed the true nature of the synthesis,[79] and whether in fact it had actually taken place, others focused on the uniquely integrative features of the synthesis to understand integration within scientific fields of study.[80] Their diverse and frequently contradictory positions with respect to the synthetic theory often added even more fuel to the evolutionary fires then raging.[81]

Then, in 1988, well after he had completed his weighty history of twentieth-century evolutionary biology, *Sewall Wright and Evolutionary Biology*,[82] William Provine proposed yet another interpretative twist to

lan, "Testability, Disreputability, and the Structure of the Modern Synthetic Theory of Evolution," *Erkenntnis* 13 (1978): 261–78. Anna Riddiford and David Penny, "The Scientific Status of Modern Evolutionary Theory," in Pollard, ed., *Evolutionary Theory*, pp. 1–38; E. A. Lloyd, *The Structure and Confirmation of Evolutionary Theory* (Westport, Conn.: Greenwood, 1988). On the structure of biology, see Alexander Rosenberg, *The Structure of Biological Science* (Cambridge: Cambridge University Press, 1985); see also David L. Hull, *The Philosophy of Biological Science* (Englewood Cliffs, N.J.: Prentice Hall, 1974); and Elliot Sober, *The Nature of Selection: Evolutionary Theory in Philosophical Focus* (Cambridge: MIT Press, 1984); idem, ed., *Conceptual Issues in Evolutionary Biology* (Cambridge: MIT Press, 1984).

[78] Richard Burian, "Challenges to the Evolutionary Synthesis," *Evolutionary Biology* 23 (1988): 247–69; G. D. Wasserman, "On the Nature of the Theory of Evolution," *Philosophy of Science* 48 (1981): 416–37; J. Tuomi, "Structure and Dynamics of Darwinian Evolutionary Theory," *Systematic Zoology* 30 (1981): 22–31; and Caplan, "Testability, Disreputability, and the Structure of the Modern Synthetic Theory of Evolution."

[79] Alexander Rosenberg, "Genetics and the Theory of Natural Selection: Synthesis or Sustinance?" *Nature and System* 1 (1979): 3–15; Mary B. Williams, "Deducing the Consequences of Evolution: A Mathematical Model," *Journal of Theoretical Biology* 29 (1970): 343–85. See also G. Van Balen, "The Darwinian Synthesis: A Critique of the Rosenberg/Williams Argument," *British Journal for the Philosophy of Science* 39 (1988): 441–48; and Marjorie Grene, "Changing Concepts of Darwinian Evolution," *The Monist* 64 (1981): 195–213.

[80] William Bechtel, ed., *Integrating Scientific Disciplines* (Dordrecht: Martinus Nijhoff, 1986). The contributions of Lindley Darden, John Beatty, and Marjorie Grene made special reference to the evolutionary synthesis: John Beatty, "The Synthesis and the Synthetic Theory," in Bechtel, ed., *Integrating Scientific Disciplines*, pp. 125–35; Lindley Darden, "Relations among the Fields in the Evolutionary Synthesis," in Bechtel, ed., *Integrating Scientific Disciplines*, pp. 113–23; and Marjorie Grene, "Introduction," in Bechtel, ed., *Integrating Scientific Disciplines*, pp. 145–48.

[81] The most insightful philosophical contribution appeared in 1990: Jean Gayon, "Critics and Criticisms of the Modern Synthesis: The Viewpoint of a Philosopher," *Evolutionary Biology* 24 (1990): 1–49. See also the review article on evolution that appeared in the same volume: Marjorie Grene, "Is Evolution at a Crossroads?" *Evolutionary Biology* 24 (1990): 51–80.

[82] William B. Provine, *Sewall Wright and Evolutionary Biology* (Chicago: University of Chicago Press, 1986); see also idem, ed., *Sewall Wright: Evolution—Selected Papers* (Chicago: University of Chicago Press, 1986).

understanding the synthesis of the 1930s and 1940s. Pointing to the fact that books of evolution decreased in size during and after the synthesis, he noted that discussions of alternative mechanisms of evolution had diminished after the 1930s. Thus, while for Provine and others it was hard to determine what exactly constituted the synthesis, it was not so hard to figure out what was thrown out: Lamarckism, saltationism/mutationism (as in de Vriesian *Mutationstheorie*), directed evolution (as in aristogenesis, orthogenesis, etc.), and/or other purported teleological causes of evolutionary change. Instead of a true synthesis of evolution, Provine instead proposed that a "constriction" of evolutionary theory had taken place as a narrowing of viable mechanisms occurred: "The evolutionary synthesis was not so much a synthesis as it was a vast cut-down of variables considered important in the evolutionary process."[83] Provine concluded with the renaming of the evolutionary synthesis as the "evolutionary constriction":

> The term 'evolutionary constriction' helps us to understand that evolutionists after 1930 might disagree intensely with each other about effective population size, population structure, random genetic drift, levels of heterozygosity, mutation rates, migration rates, etc., but all could agree that these variables were or could be important in evolution in nature, and that purposive forces played no role at all. So the agreement was on the set of variables, and the disagreement concerned differences in evaluating relative influences of the agreed-upon variables. I agree with Gould that evolutionary biology 'hardened' toward a selectionist interpretation especially during the late 1940s and 1950s. I see this as a further constriction of the evolutionary constriction (but I like the sound of 'hardening of the constriction.')[84]

By the late 1980s the notoriety of the evolutionary synthesis was recognized. The numerous critics and commentators had drawn attention to its historical significance, yet little historical understanding and consensus had been reached. If anything, the numerous critics and commentators had served to confuse and frustrate each other in the process. So notorious did "the synthesis" become, that few serious historically minded analysts would touch the subject, let alone know where to begin to sort through the interpretive mess left behind by the numerous critics and

[83] Provine, "Progress in Evolution and Meaning in Life," p. 61. A slightly revised edition appeared in 1992: William B. Provine, "Progress in Evolution and Meaning in Life," in C. Kenneth Waters and Albert Van Helden, eds., *Julian Huxley, Biologist and Statesman of Science* (Houston: Rice University Press, 1992).

[84] Ibid., p. 61.

commentators. The interval of time between 1974 and 1987 had thus seen a period of debate and controversy pertaining to evolution that had not been seen since the turn-of-the-century "eclipse" of Darwin. Under the constant shadow of controversy, the evolutionary synthesis took on more and more of an enigmatic and elusive quality; for would-be analysts, members of what had become a "Synthesis Industry" that began to rival the "Darwin Industry" in output, it became what philosopher of science Richard Burian aptly called a "moving target."[85]

[85] Burian, "Challenges to the Evolutionary Synthesis."

Rethinking the Evolutionary Synthesis: Historiographic Questions and Perspectives Explored

> The historiography of science in some ways greatly resembles the history of science herself.
>
> Ernst Mayr, *The Recent Historiography of Genetics*

> ... a philosophy is characterized more by the *formulation* of its problems than by its solution of them. Its answers establish an edifice of facts; but its questions make the frame in which its pictures of facts is plotted. They make more than the frame; they give the angle of perspective, the palette, the style in which the picture is drawn—everything except the subject. In our questions lie our *principles of analysis,* and our answers may express whatever those principles are able to yield.
>
> Susanne K. Langer, *Philosophy in a New Key*

As FRUITFUL discussions and relevant literature on the synthesis began to wane in the late 1980s, a fresh cohort of historical workers turned to what had become one of the central problems of the history of biology. Having witnessed methodological debates in the history and philosophy of science, and having familiarity with sociological methods that were making their way into the history and philosophy of science, some of these workers began to rethink the evolutionary synthesis using varied historiographic methods and models. Specifically, they discussed the comparative values of newer "external" versus the more established and conventional "internal" approaches to the history of science and whether alternative "contextualist" perspectives might shed light on the evolutionary synthesis.

RETHINKING THE EVOLUTIONARY SYNTHESIS

Historians of science as a whole had been actively exploring alternatives to internalist approaches.[1] Rather than viewing science as the simple

---

[1] For the most recent discussion on the external–internal debates in the history of science, see Steven Shapin, "Discipline and Bounding: The History and Sociology of Science

outcome of logical processes that transcended society, externalist (and/or related sociological) approaches began to dominate scholarship in the history of science. Science was no longer the practice of elite "great men" that led inexorably to the truth, but instead was a vastly more complicated practice that drew on psychological, sociological, aesthetic, and other such traditionally "extrascientific" or external determinants. Following the lead of other historians of science, historians of biology also began to explore various contexts of the history of biology.

By the late 1980s, moreover, the history of biology was itself becoming a burgeoning independent field of inquiry as the International Society for the History, Philosophy and Social Studies of Biology (ISHPSSB) was founded.[2] As the membership increased, and as they came together for conferences and meetings, historians of biology became more familiar with varied approaches and methodologies from co-workers in philosophy, sociology, and the social study of science. Though the history of biology became a vastly less well defined practice in the process, it also became enriched by the variety of historiographic perspectives that held promise for speaking to wider audiences.

To a significant extent, many of these approaches making their way into the history of biology were the result of the more diverse backgrounds of individuals entering the field. Many were increasingly receiving advanced degrees in the specific field of the history of science and philosophy of science while others were more generally trained in American history and social history—especially in contrast to earlier workers, who had been trained nearly exclusively in science. As they delved into the history of biology, they examined largely unexplored contexts of biological activity and brought fresh perspectives to the subject. Thus, by the late 1980s, institutional, organizational, national, and other such external contexts of scientific activity were successfully introduced by scholars who identified themselves as "professional" historians of science (i.e., trained extensively or nearly exclusively in the recognized academic field of the history of science). The numbers of such professional historians of science who expressed an interest in the history of the biological and related life sciences also increased. As the scholarship drawing on a range of externalist approaches to the history of science grew, so too did the

---

as Seen Through the Externalism–Internalism Debate," *History of Science* 30 (1992): 333–69.

[2] The society grew out of smaller local meetings. At the 1989 London, Ontario meetings the charter of the society was drafted. For a brief introduction to the society, see its first newsletter. See also V. B. Smocovitis, "Trends in the Recent History, Philosophy, and Social Studies of Biology," *Trends in Ecology and Evolution* 9 (1994): 4–5.

number of conferences holding sessions on related themes. Numerous edited volumes from conference proceedings appeared, many of which were exclusively devoted to lively if not provocative historiographic and methodological discussions.[3]

For historians of biology, one of the most important turning points in the field came with the publication of the volume entitled *The American Development of Biology*.[4] Edited by Ronald Rainger, Keith R. Benson, and Jane Maienschein, the volume contained the proceedings of a special conference supported by the American Society of Zoologists in celebration of their centennial anniversary (1889–1989).[5] The volume was important for the subsequent history of biology for at least two reasons. First, it was part of a "generational revolt" by younger scholars who wished to amend or revise the preceding scholarship, especially as embodied by classic texts that surveyed biology and its emergence like Garland Allen's *Life Science in the Twentieth Century*.[6] Second, the volume used a variety of historiographic approaches that either asked new questions of existing scientific material or examined previously unexplored sources. In focusing on the American context of the biological sciences, furthermore, it served to redress an imbalance in the history of biology that had stressed earlier French, English, or German contexts. The focus on the historical development of biology in America at the turn of the century and whether there could be a "character" to biological science that rested on American cultural foundations was therefore one of the important contributions of the volume to subsequent scholarship.[7] The

[3] See Sally Gregory Kohlstedt and Margaret Rossiter, eds., *Historical Writing on American Science: Perspectives and Prospects* (Baltimore: Johns Hopkins University Press, 1985). For an excellent survey of some of the literature in the history of biology at this time and an assessment of the field as a whole, see the contribution to the volume by Jane Maienschein, "History of Biology," pp. 147–62.

[4] Ronald Rainger, Keith R. Benson, and Jane Maienschein, eds., *The American Development of Biology* (Philadelphia: University of Pennsylvania Press, 1988). A second volume, also commissioned by the American Society of Zoologists, appeared in 1991. See Keith R. Benson, Jane Maienschein, and Ronald Rainger, eds., *The Expansion of American Biology* (New Brunswick, N.J.: Rutgers University Press, 1991).

[5] The American Society of Zoologists has a division of the history and philosophy of science that supports historians and philosophers of science interested in zoology. The Botanical Society of America also supports a similar historical section, as does the American Association for the Advancement of Science.

[6] Garland Allen, *Life Science in the Twentieth Century* (New York: Wiley, 1975; reprint Cambridge: Cambridge University Press, 1978). For the call to explore biology from varied perspectives, see Frederick Churchill, "In Search of the New Biology: An Epilogue," *Journal of the History of Biology* 14 (1981): 177–91. For the historical background to Churchill's call, see "Introduction," *The American Development of Biology*.

[7] The historiographic road to the history of American biology in the nineteenth century had been paved earlier by A. Hunter Dupree, *Asa Gray* (Cambridge: Cambridge University Press, 1960); Edward Lurie, *Louis Agassiz: A Life in Science* (Chicago: University of Chi-

volume generally stressed institutional, organizational, and more specific contexts or sites of biological activity, examining (for example) the fate of biology at such key American universities like the University of Chicago, Johns Hopkins University, and Harvard University, the organization of scientific societies like the newer American Institute of Biological Sciences and the older American Society of Naturalists, as well as the immediate scientific sites of activity such as museums, laboratories, and field stations. It also left enough room for discussion of intellectual matters and scientific content. Thus, though it stressed the external features of biological science, it did not do so at the expense of the internal features of biology; nor did it in any way diminish the intellectual and scientific components of science. Ironically, however, as several of the authors attempted to redress a historiographic imbalance that had favored internalist approaches and intellectual histories of biological subjects, they had also simultaneously and unwittingly supported the great divide between the two. At the same time that many of the contributors argued that both internal and external determinants needed inclusion for proper historical understanding of biology, they also supported the dichotomy or split between the two by trying to show connections and causal influences flowing back and forth.

The possibility of collapsing the distinction between internal and external determinants of science had actually become historiographic reality by the mid-1980s. As the sociology of knowledge made its way into historical circles via the publication of critical works like *Leviathan and the Air-Pump*,[8] and as philosophers of science began to move away from theory-dominated accounts of science to examinations of scientific practice (including experiments, modeling, and other such procedures),[9] discussion of contextualist historiography, which collapsed the internal–external distinction, became a lively topic for discussion. By the late 1980s the history of science was seeing not only an increase in what could be called more traditional externalist approaches and social histories of sci-

---

cago Press, 1960); and idem, *Nature and the American Mind: Louis Agassiz and the Culture of Science* (New York: Science History Publications, 1974).

[8] Steven Shapin and Simon Schaffer, *Leviathan and the Air-Pump: Hobbes, Boyle, and the Experimental Life* (Princeton: Princeton University Press, 1985). See the discussion in chap. 4.

[9] For the early philosophical literature urging a move away from theory-dominated accounts, see Nancy Cartwright, *How the Laws of Physics Lie* (Oxford: Oxford University Press, 1983); Ian Hacking, *Representing and Intervening* (Cambridge: Cambridge University Press, 1983); Peter Galison, *How Experiments End* (Chicago: University of Chicago Press, 1987); Timothy Lenoir, "Practice, Reason, Context: The Dialogue between Theory and Experiment," *Science in Context* 2 (1988): 3–22; see also Peter Galison, "History, Philosophy and the Central Metaphor," *Science in Context* 2 (1988): 197–212.

ence as a whole, but also the injection of methods and approaches that tried to remove the age-old dichotomy. Under the banner of "contextualism" (which could actually refer to a variety of aims, methods, and purposes) some began to view science not only as social activity, but to also argue for the much stronger claim that scientific knowledge was a social construct. So much had these approaches become part of the apparatus for historical discussion of science that it frequently worked to the disadvantage of traditional history of science: the occasion of the three hundreth anniversary of the publication of Newton's *Principia*, which should have been reason for pause and reflection if not celebration, went by largely unnoticed by the leading American journal for the history of science, *Isis*.[10] By 1987 the subtitle for *Isis* ("An international review devoted to the history of science and its cultural influences") weighed more and more in favor of the latter half of the phrase.

While the introduction of this social or sociological dimension met with enormous controversy and resistance, it also led to novel insights into scientific practice. Studies on scientific negotiation leading to consensus formation and on collectives of scientists and how they became organized into scientific disciplines all became popular subjects of historical inquiry.[11] Historical workers became increasingly excited by such successful sociological applications to history and turned to sociology, while sociologists turned their analytical interests and methods to scientific practice. As the numbers of workers interested in the history of science increased, and as approaches from fields like sociology joined more traditional history and to some degree philosophy, a new transdisciplinary approach calling itself "science studies" emerged. By the late 1980s some American universities began to institute such transdisciplinary approaches, which viewed the history, philosophy, and sociology of scientific knowledge as "inextricably linked."[12]

---

[10] The *British Journal for the History of Science* and the more historiographic *History of Science* journal also did not note the anniversary.

[11] Gerard Lemaine, et al., eds., *Perspectives on the Emergence of Scientific Disciplines* (The Hague: Mouton, 1976); David Edge and Michael Mulkay, *Astronomy Transformed: The Emergence of Radio Astronomy in Britain* (New York: Wiley, 1976); Robert E. Kohler, *From Medical Chemistry to Biochemistry: The Making of a Biomedical Discipline* (Cambridge: Cambridge University Press, 1982); Thomas Söderquist, *The Ecologists: From Merry Naturalists to Saviours of the Nation* (Stockholm: Almqvist and Wiksell International, 1986); and Robert Marc Friedman, *Appropriating the Weather: Vilhelm Bjerknes and the Construction of Modern Meteorology* (Ithaca: Cornell University Press, 1989).

[12] The new configuration did not escape the notice of the wider scholarly community. See Chris Raymond, "Scholars Take a New Approach in Studying the Institution of Science," *Chronicle of Higher Education*, 9 May 1990, pp. A4–A7. See the discussion of science studies in chap. 4.

Against the backdrop of efforts to understand external determinants of scientific activity and focus on the social and sociological history of science, some scholars began to call for similar approaches to the history of disciplines like ecology and evolutionary biology.[13] Curiously, unlike molecular biology—whose institutional contexts had been, and were being, actively explored—evolutionary biology, especially in its twentieth-century guise, seemed nearly completely removed from such historiographic considerations. While historians of molecular biology were busy working out patterns of funding or support, exploring ties to medical schools, pharmaceutical companies, and other industries, and investigating social patterns of the professionalization of their science,[14] historians of the more "classical" biology were either busy mining the Darwin Industry or continuing to explore the rise of classical genetics and the relative contributions of geneticists. Suitably overwhelmed by the persistent problems of the evolutionary synthesis and the sharp twists and turns taking place in contemporary evolutionary debates, historians of evolution lacked not only the resources but also the energy to even begin to think of approaching the synthesis from social or sociological directions. Thus, despite the wealth of methodological resources being developed in wider science studies circles, the historical and sociological picture of the synthesis was not only messy but remarkably crude, neglecting key historical events and factors that should otherwise have seemed obvious to its students. As a result, the fact that the synthesis had taken place during one of the greatest political and social upheavals known as the Second World War had been little noted, nor that the generation of architects had inherited the memory of the carnage of the First World War and the Great Depression.[15] Outside of the few studies on Julian Huxley (the most ob-

[13] John Beatty, "Ecology and Evolutionary Biology in the War and Postwar Years: Questions and Comments," *Journal of the History of Biology* 21 (1988): 245–63. Beatty's commentary was in response to articles by Gregg Mitman, Evelyn Fox Keller, and Peter J. Taylor. Even though it was an important event for evolutionary biologists, the evolutionary synthesis was not discussed in this suite of articles.

[14] See, for instance, the work of Robert E. Kohler, Pnina G. Abir-Am, and Lily E. Kay: Robert E. Kohler, "The Management of Science: The Experience of Warren Weaver and the Rockefeller Foundation Programme in Molecular Biology," *Minerva* 14 (1976): 279–306; idem, *From Medical Chemistry to Biochemistry: The Making of a Biomedical Discipline* (Cambridge: Cambridge University Press, 1982); Pnina G. Abir-Am, "The Assessment of Interdisciplinary Research in the 1930s: The Rockefeller Foundation and Physico-Chemical Morphology," *Minerva* 26 (1988): 153–76; and Lily E. Kay, *The Molecular Vision of Life* (New York: Oxford University Press, 1993). See also Edward J. Yoxen, "Giving Life a New Meaning: The Rise of the Molecular Biology Establishment," in N. Elias, H. Martins, and R. Whitly, eds., *Scientific Establishments and Hierarchies: Sociology of the Sciences*, vol. 4 (Dordrecht: D. Reidel, 1982), pp. 123–43.

[15] This, despite detailed study of ecology during the First World War and the interwar

vious politically and publicly engaged of the architects) no one had examined how the rise of political ideologies such as fascism, Nazism, communism, and the horrific biological nightmare of Lysenkoism had affected the architects (this despite the brutal elimination of so many geneticists in the Soviet Union).[16] So, too, philosophical movements like positivism, the movement toward a progressively secular and liberal worldview, and other cultural movements including "internationalism," "modernism," and the drive to create a unified global community, had not been examined in much detail for their relationship to the developments in evolutionary studies.[17] How the architects of the synthesis "networked" or interacted with other intellectuals who may have appeared far apart both intellectually and geographically during this interval of time had also not received proper discussion. That the community of players was actually quite small and had ready access to the growing communications technology, to the inexpensively produced and reproduced written word, and to increasingly rapid means of inexpensive travel that could bring like-minded intellectuals together was not considered. How such developments leading to large-scale cultural alterations in the growing global community affected the generation of architects had not been noted at all.

As a collective or social enterprise, moreover, evolutionary biology might as well have been the terra incognita for historians of evolution. Not only were details on the history of the discipline lacking, but its historical existence had not even been recognized. When exactly the self-identified group of practitioners adopted the appellation "evolutionary biology" was not known, nor was there knowledge of the conditions for its emergence (such as how and where it was institutionalized), let alone how (and from where) its members were enrolled and enculturated. Its major texts, tenets, critical problems, and established procedures had not even been well defined. Its relationship to closely neighboring disciplines like genetics, which had emerged at the turn of the century following the

period. See Gregg Mitman, *The State of Nature: Ecology, Community, and American Social Thought, 1900–1950* (Chicago: University of Chicago Press, 1992). This book drew on Mitman's earlier studies, which included consideration of evolution during the First World War: see Gregg Mitman, "Evolution as Gospel: William Patten, the Language of Democracy, and the Great War," *Isis* 81 (1990): 446–63.

[16] Colin Divall, "Capitalising on 'Science': Philosophical Ambiguity in Julian Huxley's Politics, 1920–1950" (Ph.D. diss., University of Manchester, 1985); and see John C. Greene, "From Huxley to Huxley: Transformations in the Darwinian Credo," in *Science, Ideology and World View* (Berkeley: University of California Press, 1981), pp. 58–193.

[17] The sole exception is an article-length exploration of the history of evolutionary ideas at this time. See John C. Greene, "The History of Ideas Revisited," *Revue de synthèse* 4 (1986): 201–27.

rediscovery of Mendel, was not critically explored, nor was the connection between evolutionary biology and what would eventually be seen as its close "rival" science, molecular biology. That molecular biology (and biochemistry) were heavily funded, while the natural history-oriented evolutionists had difficulty receiving similar funding had been historically noted by Ernst Mayr, but few others had followed up on his observations. So, too, few had followed up on, or even believed that Darwinism, evolution, and/or natural history had been undermined by the origin of more experimental disciplines like genetics.[18] Reinforced by the work of Garland Allen (known as the "Allen" thesis to younger historians of science) the "tension" between experimentalists versus naturalists occupied a significant number of historians of biology.[19] So much had it become an object of historical dispute (whether or not it took place, and/or the extent to which it operated) that no one rechecked Mayr's repeated historical observation that courses in evolutionary biology were hard to come by until the 1950s, and that leading textbooks of biology ignored or dismissed evolution until about the same time. (These would have been strong indicators of the increasing unpopularity of evolutionary studies during a critically important interval of time.) But while there clearly appeared to be some earlier criticism directed toward evolutionary science (recall that even Huxley had pointed to the "eclipse of Darwin" at the turn of the century in his 1942 book), the conditions under which evolutionary biology had been legitimated as a proper science had not been explored at all. Even two of the remaining original participants/ architects had been initially baffled by historical questions concerning the exact origins of their discipline (this despite their pronounced historical interest in evolution, which greatly exceeded the attention given to historical matters by most molecular biologists).[20] Queried directly about the historical origins of his discipline of evolutionary biology, Ernst Mayr responded, "Frankly, I do not know when the phrase evolutionary biology was first coined. In all my early work on evolution, I considered myself to be an avian systematist. I think the founding of the Society for the Study of Evolution in 1946 was certainly a crucial part of the devel-

---

[18] Reasons for this range from the rigorous and experimental nature of the science of genetics to the bandwagon effect that the new "fashionable" science of genetics had generated. Certainly the fact that genetics had applied aspects leading to medical treatments and agricultural products has a great deal to do with its generous support. Mayr has discussed the tensions between geneticists and experimental biologists and their more natural history and evolutionary colleagues in his corpus of historical work.

[19] Garland Allen, "Naturalists and Experimentalists: The Genotype and the Phenotype," *Studies in History of Biology* 3 (1979): 179–209.

[20] Gunther Stent is a notable exception.

opment. The evolutionists finally had a flag around which they could rally."[21] His contemporary G. Ledyard Stebbins, Jr. had a nearly identical (and independent) response to the same question: "I don't know when the exact term 'Evolutionary Biology' was first used. All of us who aided the launching of the journal *Evolution* during the 1940's had it very much in mind."[22]

On such historical rethinking, one possible reason that the picture of the evolutionary synthesis seemed such a mess was not solely because of the number of contradictory interpretations that had been offered, but also because in actuality, *too few* studies of the wider historical and socio-logical contexts of the evolutionary synthesis had even been considered. This appeared so much the case that even the second volume commissioned by the American Society of Zoologists dedicated to historical exploration of American biology between 1920 and 1950 devoted only the briefest of references to the evolutionary synthesis, despite the fact that the historical periods mapped a perfect chronological correspondence with the aim of the volume.[23]

Moves toward understanding the wider historical and social contexts of the synthesis began in earnest in the late 1980s. A typographical error in the title of a conference paper mistakenly reading "On the Origins of Evolutionary Biology" in 1989 induced at least one worker to finally turn to this historical question full-time.[24] The linkage between the emergence of the discipline and the conditions leading to the synthesis were explored. At the same time, historiographic questions of the synthesis were posed as a rethinking of the evolutionary synthesis came from different directions.

[21] Letter to author from Ernst Mayr, 27 February 1989. Mayr later restated this; letter to author from Ernst Mayr, 15 August 1989.

[22] Letter to author from G. Ledyard Stebbins, Jr., 4 May 1989.

[23] Keith R. Benson, Jane Maienschein, and Ronald Rainger, eds., *The Expansion of American Biology* (New Brunswick, N.J.: Rutgers University Press, 1991).

[24] The title had been sent to John Harley Warner and Frederic L. Holmes, the conference organizers, as "The Origins of Evolutionary Botany." A typographical error in the program substituted "biology" for "botany" (letter to author from John Harley Warner, 15 February 1989). See V. B. Smocovitis, "On the Origins of Evolutionary Biology," paper delivered at the Joint Atlantic Seminar in the History of Biology, Yale University, April 1989; "Rethinking the Evolutionary Synthesis," paper delivered at the Plenary Session of the International Society for the History, Philosophy and Social Studies of Biology, London, Ontario, July 1989. New approaches to the synthesis were addressed at two twin sessions on the evolutionary synthesis at the 1990 History of Science Society meetings in Seattle, Washington. Participants included Mark B. Adams, William B. Provine, and Garland Allen. The second session included Ronald Rainger, Marc Swetlitz, Joseph Allen Cain, and Léo Laporte. Abstracts of the papers appeared with the conference program. Another session organized by Elihu Gerson was held at the 1991 meetings of the International Society for the History, Philosophy, and Social Studies of Biology in Evanston, Illinois.

HISTORIOGRAPHIC QUESTIONS EXPLORED

On close consideration of the existing literature, nearly all accounts of the synthesis had stressed the establishment of the synthetic theory, attempted to sort its contents, and tried to pinpoint its cognitive components. In so doing, analysts had become bogged down by an essentialistic approach to biology and evolution that could only fail to turn up the much sought-after "core set" of beliefs of the theory. Historians and philosophers of evolution had focused so intensely on the status, validity, and structure of the synthetic theory that they had ignored other features of scientific practice that properly deserved equal ranking to theory. What sort of understanding of the synthesis would scholars gain if they additionally explored critical methods and procedures such as the use of mathematical modeling, modes and designs of experimentation, the inclusion of varied geographic environments (and surveys of populational and variational patterns), instrumentation, and the choice of model organismic systems? And, bearing close relation to the distinction between scientific practice and theory, could the "evolutionary synthesis" be effectively disengaged from the "synthetic theory," as philosopher John Beatty had insightfully suggested? What purposes would be served by making the distinction between the two? And would such a distinction lead to a clearer understanding of the synthesis? Consideration of the practical components of evolution, in addition to the cognitive, intellectual, or theoretical components, thus had great potential to give a new perspective on the problem of the synthesis. These considerations could also begin first to highlight and then to sort through the terminological confusion in the language of the synthesis: while the term "the synthetic theory" referred to the theory, "the evolutionary synthesis" could refer to the historical event designated by Mayr and Provine after their 1974 workshops, while "modern synthesis" could hold for Huxley's original appellation for what he viewed as developments in evolution around 1942 in his book *Evolution: The Modern Synthesis*. That these terms did not necessarily refer to the same thing had become apparent by the late 1980s. Historical purists quickly began to carefully denote the differences in use as "Huxley's Synthesis" began specifically to refer to the 1942 appellation.

Stressing the practical features of evolution also opened the door to other fruitful venues for exploration. Such explorations included a rethinking of the larger picture of the history of twentieth-century biology: What would this bigger picture of the history of twentieth-century evolution look like with equal time given to the practices of evolutionists? For one thing, a consideration of "practice" would open inquiry into the methods of evolutionists, which were largely unexplored by historians and philosophers who had relied more heavily on exemplars from the physical sciences. Methodological inquiry into evolution foregrounded several interesting problems. The first came as the result of the problems of dealing with a historical science that could not be observed directly and could not lend itself (at least easily) to direct interventionist manipulation such as that characterized by experimentation in the physical sciences. How would evolutionists perform experiments in such a science, and could such a science hold as a proper science in the absence of the experimental methodology that had served the physical sciences so well? When did evolutionists begin to perform experiments within such a science, and how were these experiments designed around these natural limitations? How did evolutionists interpret their data, especially in a science that drew so heavily on interpretive methodology? How did they construct their evidentiary base in support of a clearly historical process? And how would considerations of evolutionary processes as indicated by patterns in the fossil record—the traces of past processes—rather than more recent evolutionary trends alter understanding of evolution?

A second interesting methodological aspect that emerged from the focus on practice was related to the site of scientific activity, which dealt with a greater range of conditions than the physical sciences. Under what geographical conditions, for instance, were studies undertaken—in the wild on "natural" populations, or under simulated or laboratory conditions on solitary individuals? To what extent were such simulations or attempts to model natural events commensurate with "nature"? When were quantitative methods applied to evolution, under what conditions, and for what parameters? How were such studies eventually interpreted and then integrated by the varied and disparately situated members of the evolutionary community?

Explorations into the site of scientific activity also revealed the disparate institutional sites in which evolution was being studied. Here consideration of the differences among museum workers, government or other field biologists, agriculturists, and university- and industry-based laboratory scientists had the potential of bringing out differences in theoretical perspectives that emerged from individuals operating in such sites. What exactly were field biologists, museum workers, and agriculturists involved in the synthesis actually doing during the period of the synthesis and how did their activity inform their theoretical beliefs? How did the actual site of scientific activity, ranging from the laboratory, the field, the farm, the desk, or the zoo, mold and shape the perspectives of the evolutionists taking part in the synthesis? While many of the protagonists in the synthesis emerged as evolutionary biologists, all had held positions or been trained as avian systematists, paleontologists, geneticists, botanists, and the like—positions that not only drew considerable time away from theoretical evolutionary work, but also would have shaped their final theoretical beliefs and specific evolutionary concerns.

Methodological inquiry into practice also introduced consideration of the choice of specific organisms for use in evolutionary studies, a problem present in the life sciences that did not exist for the physical sciences. How did evolutionists choose model organisms for their studies? How easily and to what extent could evolutionists extrapolate phenomena observed for one group to others, or, how generalizable were observations on specific organisms? Studies of evolution using plant model organisms rather than animal systems, for instance, gave rather different understandings of evolutionary processes, as had been demonstrated by the case of purported saltatory evolution that Hugo De Vries had mistakenly interpreted for the evening primrose, *Oenothera*. One outstanding reason for the success of Thomas Hunt Morgan's school of "classical" genetics, furthermore, was due to the suitability of *Drosophila melanogaster*, an organism that demonstrated genetic changes at the level of individual differences and could therefore support the gradual process of Darwinian evolution. As demonstrated by the different patterns of evolution discerned in just these two organisms, choice of study organism could therefore determine the kinds of evolutionary theories one supported or rejected.

Last of all: How did the actual fields coming to synthesis deal with their varied practices? Clearly the practice of avian systematics was different from paleontology, from genetics, from botany, and from other such well-established disciplines. How did an integration and coordination among these varied disciplines of the biological sciences take place so that participants could claim that a synthesis among evolutionary disciplines had occurred? What process of communication took place among disciplines, and how did the communication networks assemble?

Questions such as these made it all too apparent that the historiography of twentieth-century evolution had been skewed by the almost exclusive emphasis on the theory rather than the practice of evolution. Aside from Provine's work on Sewall Wright and Theodosius Dobzhansky, few had worked on the details of the collaboration between theorists and practitioners.[25] Historical examination of the much neglected other half of evolution clearly had the potential to alter understanding of the history of twentieth-century evolution significantly.

William B. Provine, for instance, had organized his massive history of twentieth-century biology, *Sewall Wright and Evolutionary Biology*, around the biographical subject of Wright because Provine was focusing his efforts to understand the history of evolutionary biology in terms of its theoretical developments, especially in mathematical and population genetic terms.[26] Equal consideration given to the more practical dimensions of evolution might have led to a different project. Consideration of fieldwork or studies of evolution in natural populations would have shifted the focus of attention to Theodosius Dobzhansky instead of Sewall Wright, for instance. A similar practical consideration of the origins of the discipline in terms of administrative and organizational structures assembled would have shifted the emphasis toward Ernst Mayr, while consideration of the public relations of evolutionists would shift the emphasis not only to the obvious contributions of Huxley, but also to G. G. Simpson and the other architects who, in addition to practicing science and writing synthetic accounts of their fields, were also popular science writers. One could just as easily have chosen any of these as biographical subjects of the history of evolutionary biology instead of, or in addition to, Sewall Wright.

Other features might have been highlighted by a greater consideration of practical matters. If theory and practice worked hand in hand or could be seen to be in dialogue with each other, then the collaboration between "theorist" and "practitioner" would have received more historical attention. The history of evolutionary biology could then be framed around the famous Wright and Dobzhansky or Fisher and Ford collaboration. The nature of this collaboration could then open fruitful discussion into how such zones of exchange and communication open up between different groups or cultures of scientists.[27]

[25] Robert Kohler has recently explored such practical considerations with some success. Although his project focuses on the classical genetics of the Morgan school, it includes discussion on evolutionary subjects, especially the work of Dobzhansky. See the discussion on Dobzhansky's choice of model organism and how it affected the "moral economy" of laboratory practice: Robert E. Kohler, *Lords of the Fly: Drosophila Genetics and the Experimental Life* (Chicago: University of Chicago Press, 1994). See also Adele E. Clarke and Joan H. Fujimura, *The Right Tools for the Job: At Work in Twentieth Century Life Sciences* (Princeton: Princeton University Press, 1992).

[26] Wright's reputation was as a "theorist," but Provine's study revealed how Wright's theoretical models closely mirrored his early insights on the practical breeding of domestic livestock.

[27] Peter Galison has termed these moments "trading zones." See his full discussion in

If the genre of the biographical subject or subjects were not favored, and the genre of disciplinary history selected instead, then one could have approached the history of twentieth-century evolution even more directly by writing the history of the community and the conditions under which it was founded. With a simple emphasis on practice, all of these could have become historiographic possibilities that could function as viable alternatives or supplements to existing approaches.

Another closely related problem stemming from the emphasis on theory was the almost exclusive emphasis on the internal features of the discipline. What, then, were the external parameters for the development of the discipline? In keeping with a sociological perspective, moreover, how exactly did such a disparate group arrive at consensus (assuming that consensus had even been first reached and then maintained at all)? Why should communities that could be effectively viewed as independent cultures with their own cultural apparatus, including their own discourse or common language, bother to speak across their cultures? How did scientific consensus come about, and what were the details of consensus formation with respect to points of agreement in evolutionary theory?

This latter consideration also dovetailed with the discussion of what could count as practice, for practice held varied meanings of its own (a fact borne out by the diversity of literature in science studies using this term).[28] The bulk of considerations of practice (including the methodological, procedural, geographic, institutional, and disciplinary aspects discussed above) did not adequately deal with discursive, textual, or literary practices of science. This was especially important to evolution, a historical science that relied little on material or instrumental technology, but used literary or narrative modes of explanation more readily.[29] How

---

Image and Logic (forthcoming). See also Steve Fuller, *Philosophy, Rhetoric, and the End of Knowledge: The Coming of Science and Technology Studies* (Madison: University of Wisconsin Press, 1993). Considerations of the power relations underlying the Wright–Dobzhansky interaction would also make for an especially interesting gender study: see Vassiliki Betty Smocovitis, "Gendering Evolutionary Biology: The Theory/Practice Dichotomy in American Evolutionary Biology (1930–1950)," paper delivered at the American Studies Association meetings, Boston, Massachusetts, 1993.

[28] See the discussion on science studies and science as discourse and practice in chap. 4.

[29] This has not escaped the attention of scholars applying various forms of narrative theory to understand evolution: T. A. Goudge, *The Ascent of Life: A Philosophical Study of the Theory of Evolution* (Toronto: University of Toronto Press, 1961); Ernst Mayr, *The Growth of Biological Thought* (Cambridge, Mass.: Harvard University Press, 1982); Greg Myers, *Writing Biology: Texts in the Social Construction of Scientific Knowledge* (Madison: University of Wisconsin Press, 1990); David Hull, *The Metaphysics of Evolution* (Albany: State University of New York Press, 1989); Misia Landau, *Narratives of Human Evolution* (New Haven: Yale University Press, 1991); Robert J. O'Hara, "Telling the Tree: Narrative Representation and the Study of Evolutionary History," *Biology and Philosophy* 7 (1992):

did the interplay of discursive and nondiscursive components of science (assuming that the two were distinct) create coherent narratives of evolution?[30] In keeping with the considerations of literary practice, the specific writings of the synthesis needed close examination, especially because so many of the key events in the synthesis involved the appearance of key works. Subsequent writings of the architects intended for varied audiences also needed inclusion in the story of the synthesis: nearly all of the protagonists were prolific writers who wrote not only scientific works, articles, and reviews, but who also wrote widely for semipopular and popular audiences. Many of the architects had taken advantage of the new communications technology and had extended their spheres of influence to radio and television audiences eager to learn of contemporary developments in science.

Questions such as these that drew even more attention to the inadequacies of existing accounts of the synthesis made it apparent that all accounts had focused on explaining not the reasons for consent, but the reasons for dissent preceding the synthesis. At least two workers had suggested that the evolutionary synthesis was best viewed as a "treaty" for this reason.[31] All historical explanations had focused subsequently on the "removal of these barriers" so that the proper path toward true scientific knowledge had been cleared of such unnecessary obstructions to scientific progress. Thus, the emphasis on explaining reasons for dissent rather than reasons for consent was concomitant with the philosophical commitment that held that science was a linear, progressive activity leading inexorably to truth. Within this classic view, truthful (or, more precisely, the most truthful) claims were made so that a more truthful (and better) account of the world ultimately prevailed, regardless of obstacles, hindrances, or mistakes (hence the oft-heard phrase that science is a self-correcting process).

In the specific case of the evolutionary synthesis, historical efforts had focused on explaining why geneticists had been alienated from naturalists—in "two camps" or as rival factions. In this view, the synthetic perspective arose naturally once protagonists died, taking with them their personal animosities, and once the mathematical models were developed that demonstrated that selection theory and modern genetics could be

---

135–60; and Robert J. Richards, "The Structure of Narrative Explanation," in Matthew H. Nitecki and Doris V. Nitecki, eds., *History and Evolution* (Albany: State University of New York Press, 1992).

[30] See the discussion in chap. 4.

[31] See David J. Depew and Bruce H. Weber, eds. *Evolution at a Crossroads: The New Biology and Philosophy of Science* (Cambridge: MIT Press, 1985).

effectively integrated within a populational—and geographical—frame-work through the work of systematist-naturalist-geneticists like Theodosius Dobzhansky in whose person the synthesis could take place unheeded. Because of such underlying philosophical commitments (or assumptions) of the philosophy of science, the turn-of-the-century debates in evolutionary theory had been viewed as "setting back" or "delaying" the evolutionary synthesis.[32]

Although scientists continued their work with the philosophical belief that they were engaged in a self-correcting process that ultimately led to truthful claims,[33] philosophers and historians of science had increasingly challenged the same philosophy, especially following the introduction of sociological approaches and the dissemination of Kuhn's *Structure of Scientific Revolutions*. By the late 1980s various philosophical positions had been presented. Whereas philosophers of science reached little agreement on the "nature" of the scientific enterprise, they had nearly all agreed that the cumulative growth model needed heavy emendations, if not a complete overhaul.[34] What would the evolutionary synthesis look like in light of these philosophical emendations questioning the scientific enterprise and viewing it as a culturally embedded and historically rooted practice? For one thing, questions would explore reasons for the emergence of the collective and the social reconfiguration around the discipline of evolution. More specifically, questions could then be reframed around what pulled the discipline together, rather than just explaining why individual portions were kept apart. Why did geneticists begin to engage naturalist-systematists, paleontologists, botanists, and others about evolution? How, then, did a coherent community drawing on such disparate disciplines coalesce?

Further rethinking of historiographic matters (more precisely, the philosophy of science held by historians of science) also drew attention to another feature common to all approaches to the synthesis: all had been either written by participants, scientists-turned-historians of science, or others, nearly all of whom had held to a similar philosophy of science. That the writing of histories serves different uses and purposes—long

---

[32] Both Mayr and Provine would agree on these interpretive features. See Ernst Mayr, *The Growth of Biological Thought* (Cambridge, Mass.: Harvard University Press, 1982); William Provine, *The Origins of Theoretical Population Genetics* (Chicago: University of Chicago Press, 1971). This is also adopted by Peter Bowler in *Evolution: The History of an Idea* (Berkeley: University of California Press, 1983; rev. ed. 1989).

[33] This same philosophy makes it possible for scientists to engage in their scientific work.

[34] See Larry Laudan, *Progress and Its Problems: Toward a Theory of Scientific Growth* (Berkeley: University of California Press, 1977).

noted by historical thinkers like Nietzsche—had not been fully recognized by historians of the synthesis who had not yet processed, let alone questioned, the meaning of historiography as the *writing of history*.

Such self-critical historical examination would have been especially important in the case of the evolutionary synthesis. If consensus and discipline formation were going to be part of the history of the synthesis, then how the community constructed its historical origins and how the history of the group shaped the identities of its members also needed close analytic examination. That such "disciplinary" histories served many purposes, among which were included the legitimation of the discipline, its promotion to others, the definition of inclusion—exclusion criteria (as in the gatekeeping functions of defining the insiders and outsiders of the community), and general delineation and demarcation of the boundaries of the discipline in question, was increasingly recognized by historiographers and sociologists of science.[35]

What would such historiographically informed analyses of the synthesis look like? What was the relationship of the historical event and its subsequent historical representation(s)? To what extent had the participants actively rewritten their history at the same time that they were retelling it? In the case of evolutionary biology these were especially important considerations because the writing of the history of evolution and the theory/practice of evolution were so naturally blurred. With protagonists like Ernst Mayr and Stephen Jay Gould, who were rewriting their personal histories along with their disciplinary histories, and at the same time rewriting evolutionary theory and practice, it was practically impossible to separate the theory and practice of their science from their historical interpretations.

As a profoundly historical and philosophical discipline, furthermore, evolutionary biology had more than its share of such protagonists who naturally blurred their philosophy of science with their philosophy of history.[36] These protagonists were also so sensitive to their own history, and their constant reworking of it, that they actively sought to preserve historical documentation (recall, for instance, the detailed transcripts of the 1974 evolutionary synthesis workshops and the goal of recording and then making material easily available for subsequent workers—all a

---

[35] Loren Graham, Wolf Lepenies, and Peter Weingart, eds., *Functions and Uses of Disciplinary Histories* (Dordrecht: D. Reidel, 1983).

[36] The connections between history and evolution were discussed at the 1989 Spring Systematics Symposium at the Chicago Field Museum of Natural History. See Nitecki and Nitecki, eds., *History and Evolution*.

critical part of the conference as orchestrated by Mayr).[37] The historical interests of evolutionists (though this has never been studied rigorously) far exceeded the historical interests and engagement of other scientific groups. The case of the evolutionary synthesis therefore raised an especially intricate set of issues with respect to the historiography of science; this was an especially fascinating instance of a scientific theory that appeared to be revised and rewritten in tandem with the historical and philosophical reflections of the protagonists. No historical account was removed from the scientific perspectives of its writers, and no account of evolutionary theory was removed from its historical packaging.

While all these historiographic dimensions needed constant, close attention, equal time had to be given to the actual writing of a viable historical account of the "event" in question. How would all these considerations help the historian proceed with such an account? Where would they begin to reframe a workable history of the synthesis? For a beginning, historians could focus their historical vision on the community of scientists rather than on the contributions of individuals (or, rather, question what counted as a "contribution," for this clearly played into the collective entity that served to recognize or ultimately determine the contribution). Rather than focus on the contributions of individuals, the historical vision might focus instead on the community of scientists, with the discipline as the historical unit of analysis. Thus, rather than trying to extrapolate larger patterns from the individuals to understand the community as whole (as in adding up the work of individuals to calculate the sum of the work of the collective), the community might serve directly as the unit of analysis, a unit that could be seen to override or transcend even the additive effects of individual activity.

Because this latter point is wrapped up with belief in the transcendence of scientific knowledge, the choice of community as historical unit posed a series of difficult, possibly even distasteful problems for philosophically minded historians. That disciplines, like other social entities or collectives, could possibly have lives of their own disengaged from the individual members was a troubling proposition to contemplate. Agency, free will, self-expression, self-determination, and other such longstanding demonstrations of the power of the individual to mold and shape life-experience would come into question in such a powerfully social framework. Yet while such a collective enterprise that ran the danger of extinguishing the individual was (and continues to be) a troubling proposition, it *is* largely

---

[37] This was noted in chap. 2.

taken for granted by scientists, who uphold the belief that science ulti-mately "transcends" individual experience as it does culture and history. The traditional grand narrative of science, though relying on the actions of great heroic figures, also stressed (and continues to stress) the self-correcting nature of science, as though science had a life of its own disen-gaged, ultimately, from the fallibilities and sensibilities of its practitioners (as well as being essentialized). Clearly, the discipline or collective as the historical unit of analysis raised a series of large-scale and deep philo-sophical problems, but so too did it hold the promise of providing an-other vantage point or perspective from which to view the "moving tar-get" of the evolutionary synthesis. It was also a position from which it might be possible to retell a story commensurate with the perspective of the scientists who believed themselves to be in a collective enterprise seeking transcendence.

From the vantage point of the community or social level of historical analysis, the evolutionary synthesis appeared to map a perfect corre-spondence with critical features in the history of the discipline of evolu-tionary biology. If one could view the formation of disciplines in terms of the creation of textbooks, older problems being resolved with newer ones being created, new knowledge societies being founded to organize the discipline, public forums such as conferences and journals for the dissemination of information being created, rituals and other celebratory rites being instituted, along with a common language or shared discourse spoken, then the interval of time that marked the evolutionary synthesis also underscored an important moment in the emergence of the collec-tive and profoundly *cultural* enterprise called "evolutionary biology."[38]

Etymological research into the term "evolutionary biology" had also simultaneously reinforced this possibility. Though the term had been coined in a rather obscure context in the late nineteenth century and had gained very limited use in the early twentieth century, it had not gained widespread acceptance until the 1950s.[39] Shortly after this time, evolu-tionists began to reidentify themselves not just as geneticists, or bota-nists, or systematists, but also as evolutionary biologists, if the context

---

[38] Taken as a whole, this list comprises the "culture" of the scientific community.

[39] The first recorded use of the phrase "evolutionary biology" (more precisely, "evolu-tionary *biologist*") appears to have been in 1881 in a passage from naturalist Grant Allen's *Vignettes from Nature*. There also follows a 1912 use of the term in Arthur Dendy's *Outlines of Evolutionary Biology* and then a 1938 use by Gavin de Beer in *Essays on Aspects of Evolution-ary Biology Presented to E. S. Goodrich* (Oxford: Oxford University Press, 1938). See chap. 5 for more details.

required the evocation of the collective identity.[40] Thus, at the same time that the modern synthesis of evolution was being hailed, the discipline of evolutionary biology was itself actively coalescing. (Recall that Mayr and Stebbins would date the origin of their discipline around 1947, at the time of the celebratory Princeton meetings.)

Close examination of evolutionary biology also revealed the fact that the first official international scientific knowledge society for the study of evolution in the United States was officially founded during this same time. Beginning with more local organizations founded by systematists like the Society for the Study of Speciation in the late 1930s and early 1940s, then growing to encompass paleontologists and geneticists with the National Research Council-backed Committee on Common Problems in Genetics, Paleontology and Systematics, a formal society that drew evolutionists together was founded under the title of the Society for the Study of Evolution (SSE) with the aid of a small grant from the American Philosophical Society. At the same time, a forum for publication was founded, the journal *Evolution*, which would serve as "one voice" not only for the new scientific society in the United States but also for an international audience of evolutionists.[41] Study of the organization of evolution revealed that the SSE had been founded roughly during the postwar period of institutional growth that had also supported the founding of other scientific societies, the most impressive of which was the first "umbrella-like" organization for biologists, the American Institute of Biological Sciences (AIBS). The relationship of the smaller SSE to the larger AIBS became an important question for further exploration. The SSE's relations to older existing and closely related societies like the

[40] There is some evidence that paleontologists did not accept the appellation of evolutionary biology as readily. The fact that their educational and departmental affiliations were to geology rather than biology most likely contributed to their disaffection with the biological epithet. Interestingly, one of the most recent commonly used appellations is "evolutionary science." This is a much more inclusive category that can include astronomers, chemists, cytologists, social anthropologists, and even linguists. See the "Research News" column by Ann Gibbons, "Evolving Similarities—Between Disciplines," *Science* 250 (1990): 504–6. This was a report of a conference held at Cold Spring Harbor, 24–27 September 1990, entitled "Evolution: Molecules to Culture." Reporting on the conference, Gibbons wrote that after a "whirlwind tour" by Richard Dawkins and Jared Diamond, Dawkins stated that "most attendees concluded that there is a 'general science of evolution.'" Quotation on p. 504. Even Douglas J. Futuyma, the author of the leading textbook of evolutionary biology, recently used the appellation of "evolutionary science" in an article meant to be a recent justification of evolutionary biology. Douglas J. Futuyma, "The Uses of Evolutionary Biology," *Science* 267 (1995): 41–42.

[41] How the architects tried to construct "one voice" (in their own exact terms) from their cacophony of evolutionary perspectives is discussed in Vassiliki Betty Smocovitis, "Organizing Evolution: Founding the Society for the Study of Evolution 1939–1947," *Journal of the History of Biology* 7 (1994): 241–309.

American Society of Naturalists also became worthy of study, as the two negotiated potential conflicts or overlaps in their interests.

Little was also known about the financial support for the founding and maintenance of the new society for the study of evolution. How did members convince funding agencies of their worthiness? How was the new society managed, how were its institutional structures assembled, and who became its leaders? These and other questions dealing with organizational and administrative sides to the history of both evolution and biology came to be recognized.[42] Last, the connection between evolution and biology was noted: How did the widespread belief that evolution "unifies" biology come about? Was there truth to this claim, especially given the fact that some relevant disciplines like physiology and developmental biology never felt part of the evolutionary synthesis and still held some animosity toward evolutionists? And when did the belief that evolution serves as the "central organizing principle" of biology gain widespread acceptance and use?

With just a few answers to questions such a fresh perspective or vantage point to view the synthesis became a possibility. By shifting the unit of historical analysis to the level of the discipline, for example, one could bring varied components into different alignment, making possible the reweaving of alternative historical narratives. But though a new range of narratives became a possibility, not all historical narratives would do; not all, for instance, would serve to retell a story that reconstructed events in a manner felicitous or even recognizable to the perspectives of the historical actors.

HISTORIOGRAPHIC PERSPECTIVES EXPLORED

Because the narrative shifted to the origin of the discipline, the way that the discipline was "legitimated" and what this legitimation exactly involved became central questions for study. Thus, diverse "contexts of legitimation" had to be explored. Assuming, of course, that a synthesis of evolution had taken place (I will discuss this further below) contexts of legitimation could fall immediately into the two obvious categories of internal or external contexts of legitimation. In response to the question of what pulled the discipline or group together, the internalist would stress the logic of the science and would retell the standard story of the

[42] Smocovitis, "Organizing Evolution"; and Joseph Allen Cain, "Common Problems and Cooperative Solutions: Organizational Activity in Evolutionary Studies, 1936–1947," *Isis* 84 (1993): 1–25.

architects sorting out their differences to get to the truthful account of evolution. This would then "remove the barriers" so that the synthesis could take place, which in turn led to the formation of the discipline complete with society, textbooks, and so on.

Responding to the same question of what pulled the group together, the externalist would not look to the logic of the science, but would instead seek external determinants or contributing "influences." What extrascientific or social factors had come to play in "legitimating" the collective enterprise? Here lay an interpretive twist not previously appreciated that looked to external motivators that pushed the group together (rather than just the logic of the science). One possibility would be to examine what "interests" the group held in common that could be safeguarded by the collective. Close examination of such interests that the group held revealed that evolutionary studies were becoming increasingly endangered with the spread of experimental sciences, which had delegitimated the natural history-oriented evolutionary studies.

Picking up on these social interests, a summary of the story would then stress the "threat" that evolutionary studies were facing before the synthesis (as in the "eclipse of Darwin") and examine the social or possibly institutional contexts of evolutionary practice. That evolution, especially in the United States, was endangered by antievolution and pro-religious sentiment has long been noted. By 1925 the notorious "Butler Act" had been passed, banning the teaching of evolution in Tennessee schools, leading to the prosecution of John Thomas Scopes in the famous "Monkey trial."[43] With the rise of experimental sciences like molecular biology and biochemistry joining the older genetics,[44] natural history-oriented sciences were about to feel an even bigger "threat" of being undermined in the late 1930s and 1940s. A focus on institutional developments for naturalist-systematists in the 1920s and 1930s would additionally support the belief that evolutionary studies were endangered. Though the actual numbers of naturalists was peaking by the 1930s, the number of naturalists holding on to positions in university departments was actually decreasing. This was especially true for botanists, who were

[43] For the most recent account of the creationist–evolutionist controversies in the United States, see George E. Webb, *The Creationist Controversy in America* (Lexington: University Press of Kentucky, 1994).

[44] The term "molecular biology" was first used by Warren Weaver in 1938 as part of the annual report of the Rockefeller Foundation. In 1970 Weaver wrote a letter to *Science* outlining the historical origins of the term: Warren Weaver, "Molecular Biology: Origin of the Term," *Science* 170 (1970): 581–82. Molecular biology, biochemistry, and related biomedical sciences received funds from foundations like the Rockefeller Foundation.

being "engulfed" by zoologists. As academic departments in the United States began to collapse older "zoology" and "botany" departments into more general "biology" departments in the 1920s and 1930s, the same departments also began to favor increasingly appointments and research areas that stressed less and less organismic and naturalistic methodologies in favor of newer biological (and that mostly meant physiological or biochemical and other such experimental) approaches.[45] Squeezed increasingly out of departments of "biology," naturalists did, however, flourish at museums of natural history in the United States, which also underwent a period of growth at this time.[46] Though such naturalists held positions that permitted research, a significant amount of time also had to be devoted to service-related activity such as routine identification, classification, or curatorial work that left less time, personnel, and resources for more non-service oriented theoretical research. Viewed as "workers" or the rank and file of biology, such museum-based individuals were increasingly squeezed out of high-powered research institutions, positions, and grants. That the Rockefeller Foundation supported the new molecular biology and the biomedical sciences, but not evolution and natural history, had been a sore point for evolutionists. Under such increasing pressure, the workers "organized" themselves, much as labor unions did at this time around the "rhetoric" of "common ground," "cooperation," and "unity". To sum, much along the same lines as labor unions organized and as political interest groups formed, so too did disciplines of science like evolutionary biology bring together workers whose interests were safeguarded by the collective entity.[47] Following the consolidation of the "union," there also came a media campaign, with suitable accompanying "rhetoric," all of which served to convince the wider audience of scientists that the evolutionary synthesis had taken place, and that evolution was a proper science that could rival physics and chemistry. Rituals of celebration, textbooks, museum exhibits, and televised programs could all be seen as part of a well-planned "strategy"

[45] For a statistical study of the fate of botany in departments of biology, see M. R. Bolick, "Botany Departments vs. Biology Departments: Is There a Difference for Botanical Society of America Members?" *Plant Science Bulletin* 35 (1989): 2–3.

[46] For a historical analysis of the founding and management of natural history museums in the United States at this time, see Ronald Rainger, *An Agenda for Antiquity: Henry Fairfield Osborn and Vertebrate Paleontology at the American Museum of Natural History, 1890–1935* (Tuscaloosa: The University of Alabama Press, 1991). For a different historical perspective on American museums of natural history, see Donna Haraway, *Primate Visions: Gender, Race, and Nature in the World of Modern Science* (New York: Routledge, 1989).

[47] For an example of recent U.S. labor history, see Robert H. Zieger, *The CIO 1935–1955* (Chapel Hill: University of North Carolina Press, 1995).

to "sell" the science of evolution. In return, evolutionists would then receive patronage and support from government and private agencies, have access to material and other resources, and gain both legitimacy and prestige with respect to other sciences.[48]

This was an appealing narrative for a number of reasons: such "threat" models and special "interest" models for the rise of scientific disciplines, collectives, and political factions had gained favor among political historians, social historians, and somewhat paranoid critics of scientific culture. Yet the case of evolutionary biology posed a special set of problems that could not be addressed in any simple "externalist" interpretation. Central among these was the fact that the discipline never reached the legitimate administrative entity of the university department in the United States, which, after the Second World War was increasingly becoming the site of action for evolutionary activity. If it did enter within the proper title of the department, evolutionary biology was always appended to ecology and behavior or systematics, as in "ecology and evolutionary biology," "ecology, evolution, and behavior," or "ecology and systematics." Here it should be noted that despite systematists', and some ecologists' frequent arguments to the contrary, evolution is *not* synonymous with, nor a term covered inclusively by, the categories of systematics and ecology. Not all evolutionists are interested in the same problems of classification that preoccupy systematists, and some ecologists, especially those in systems ecology, at times appear at odds with evolutionary biologists. Other areas that could be accommodated by evolutionary biology, such as mathematical population genetics, theoretical population genetics, or even behavioral ecology, might also be a difficult fit under any simple category of systematics. Furthermore, the study of evolution at biochemical or chemical or cellular levels would not always be accommodated by the rubric of either systematics or ecology. Such "semantic" problems, it should be noted here, are not insignificant, as they involve problems of meaning generating disciplinary "identity crises" that frequently bring into relief hidden or underlying commitments shared by the members that lend coherence to the community in question. The seemingly trivial choice of names of the SSE and journal *Evolution* called into question the goals and purposes of the new society in the mid-1940s. A more recent dispute over the renaming of the journal of the Botanical Society of America, from the *American Journal of*

---

[48] An analogous explanatory scheme was applied to systematics: Keith Vernon, "Desperately Seeking Status: Evolutionary Systematics and the Taxonomists' Search for Respectability 1940–60," *British Journal for the History of Science* 26 (1993): 207–27.

Botany to the *American Journal of Plant Biology*, brought lively if not heated discussion over the sometimes difficult relations among botany, zoology, and biology as well as the differences in meaning among botany, plant biology, and plant science.[49] Such debates on the history of naming the organizational and administrative structures of the biological sciences have thus proven especially revelatory episodes in understanding the fundamental identifying values or markers of members of a community.

More evidence supported the curious and problematic administrative status of evolutionary biology. When Paul Weiss, founder of the Biology Council, restructured the biological sciences in the 1950s he did not include evolutionary biology within his categories of the biological sciences (though he included cellular, behavioral, molecular, and other such descriptives). Even more of a problem to accepting this historical account was the fact that granting agencies like the National Science Foundation, which began to administer sponsored research by the mid-1950s to the life sciences, did not support (and still does not support) "evolutionary biology" under the official categories of the life sciences, or even under "pertinent disciplines."[50] Reasons for this absence include the thorny problem of funding what some activist groups would consider "anticreationist" research. Nor is the category "evolutionary biology" to be seen in the numerous career pamphlets that were designed to draw young Americans to the burgeoning biological sciences in the 1960s.[51]

Given these considerations, one could of course question whether or not a proper discipline formed; yet as far as textbooks, courses of instruction, knowledge societies, journals, and self-identified appellations went, evolutionary biology could claim that it had attained disciplinary status. With respect to the scholarly sociological literature available on discipline formation, furthermore, evolutionary biology clearly defied analogous

[49] V. B. Smocovitis, "Disciplining Botany: A Taxonomic Problem," *Taxon* 41 (1992): 459–69.

[50] In 1956 the definitions of life sciences categories for federal grants and contracts for unclassified research included molecular biology, regulatory biology, structural biology, genetic biology, developmental biology, environmental biology, systematic biology, pathology, diagnosis, therapy, community health, plant management, animal management, soil management, technology, methodology, equipment design, training, scientific information, and facilities. See Federal Grants and Contracts for Unclassified Research in the Life Sciences, Fiscal Year 1956, National Science Foundation, Washington, D.C. Study of evolution fell under the category of systematic biology, which listed "pertinent disciplines" as taxonomy, paleontology, phylogeny, zoo- and phytogeography, and natural history. For historical background, see also Daniel J. Kevles, "The National Science Foundation and the Debate over Postwar Policy, 1942–1945," *Isis* 68 (1977): 5–26; and Toby Appel's forthcoming history of the National Science Foundation. The NSF was established in 1950.

[51] William W. Fox, *Careers in the Biological Sciences* (New York: Henry Z. Walck, 1963).

legitimation patterns seen in astronomy, biochemistry, molecular biology, all of which had been studied extensively by sociologists and historians of science.[52] Evolutionary biology had different, possibly less well-defined contexts of legitimation than any other discipline of knowledge. What were those contexts of legitimation? How many operated, and how could one bring these together within a coherent narrative?

In keeping with the collapsing of the distinction between internal and external determinants of science (and, therefore, internal and external legitimation), furthermore, what would such a disciplinary history of evolutionary biology taking these contexts into account then look like? And what did the sociology of knowledge have to contribute to the historiography of the synthesis? By the late 1980s problems with historiographic perspectives were actively discussed by scholars as approaches within science studies made their way into study of biology. But with due consideration, it was clear that not all approaches made it possible to tell a story or give an account of science felicitous or sympathetic to the perspective of the scientist. In fact, versions of contextualism that were being accepted by historians were leading more and more to narratives that were alien, antithetical, or unrecognizable to the scientists' understandings of their own histories. Few scientists, for instance, would feel satisfied with the simplistic social interest, threat, or overly political externally driven model of science; and the unreflective, irresponsible historical overuse of metaphorical terms like "strategy," "selling" science, and "rhetoric" (as in its crudest meaning of "mere") would hardly capture a world full of the deep philosophy, aesthetics, and meaning that the "architects" (their own self-designated term) tried to design for themselves. More immediately, and despite the tendency to use the language of economics, competition and conflict, or militarism in the science of evolution,[53] the "architects" in either public or private documents did not use the language of "threat," "interest," or "strategies" (these terms are more prevalent in contemporary scholarship than past science). As one study offering a detailed historical reconstruction based on close reading of archival documents on efforts to organize evolutionists demonstrated, organizing evolution was a well defined "dream"—in their own terms—held by those evolutionists who strove to reach a consensus and actively worked toward building "common ground."[54] This is not to say that the

[52] The literature on discipline formation is vast (see n. 11).

[53] Even Dobzhansky had relied on somewhat conflictual metaphors for evolution in his 1937 *Genetics and the Origin of Species* as did Darwin in *Origin* and other scientific work.

[54] Smocovitis, "Organizing Evolution." Similarly, Joseph Allen Cain entitled his analysis

architects were driven solely by aesthetic sensibilities devoid of social or personal "interests," but that the former played a more prominent (and possibly most prominent) role in their science than had been previously noted.

Worse still was the growing "talk" among scholars (fortunately little of which has made its way into print at this time) that the synthesis was a "rhetorical strategy" (implying empty or fallacious or with the intent of "hoodwinking" audiences) with little or no scientific substance to support its claims. According to this "rhetorical" interpretive twist, the architects had conspired to create a science in order to carve out a niche for themselves. To achieve this end, they had invented, possibly even conspired, to create a fallacious "myth" of synthesis to legitimate and promote their field and themselves to others. While explorations into rhetoric, legitimation, myth making, and historical storytelling at the collective level of the scientific discipline all held the possibility of providing fruitful (and critical) venues to understanding the evolutionary synthesis, these very same terms were subject to such wide-ranging and fluctuating meanings that they inevitably led to vastly different stories. Depending on how one thought of myth, rhetoric, argument, and persuasion, and whose form of literary or critical or social or cultural theoretical assumptions one followed deliberately or inadvertently, one could retell a vastly different range of stories, some of which appeared to resemble each other closely (depending on the point of view of the observer), but not all of which spoke meaningfully to audiences. Rethinking the evolutionary synthesis had made it possible to gain new perspectives leading to new narratives, but it was clear that not all stories were equally satisfying.

Even more historiographic perspectives became a possibility in the late 1980s and early 1990s as more varied versions of "contextualism" began to make their way into the study of science from different directions. With them, another reconfiguration that drew on anthropological, literary, and cultural models rather than just sociology made its appearance. As science emerged as a culture of its own—complete with the cultural apparatus of language, rituals, and cosmologies—the cultural study of scientific knowledge became a possibility. "Cultural studies," a transdisciplinary area of research drawing on ethnography and other forms of cultural theory that specifically problematized power relations between ethnographer/historiographer and the historical or scientific culture ex-

---

of organizational activity in evolutionary studies with two phrases that repeated themselves in the documents: "common problems" and "cooperative solutions" (Cain, "Common Problems and Cooperative Solutions").

amined, was making its way into science studies, and not without some turbulence from more established sociologists committed to science studies and the older sociology of scientific knowledge (SSK).[55] Would the cultural study of scientific knowledge, with its emphasis on science as a cultural practice and on science as discursive activity, help inform historians of the synthesis? Could a multicultural, polyphonic history be written that took into account simultaneously the voices of the historical actors and the narrator's voice all within one coherent story? What exactly would constitute a "coherent story"? And were there additional problems inherent with all such "coherent stories"?

[55] Joe Rouse, "What Are Cultural Studies of Scientific Knowledge?" *Configurations* 1 (1993): 1–22. See the rather critical comments on Rouse's article by two strong British adherents of the sociology of scientific knowledge (SSK), Harry Collins and Peter Dear: H. M. Collins, "Review of *We Have Never Been Modern*, by Bruno Latour," *Isis* 85 (1994): 672–74; Peter Dear, "Cultural History of Science: An Overview with Reflections," *Science, Technology, and Human Values* 20 (1995): 150–70.

CHAPTER FOUR

The New Contextualism: Science as Discourse and Culture

> With fictions we investigate, perhaps invent, the meaning of human life.
>
> J. Hillis Miller, *Critical Terms for Literary Study*

> I began with the desire to speak with the dead.
>
> Stephen Greenblatt, *Shakespearean Negotiations*

THE CONTEXTUAL TURN

By the early 1990s humanistic fields as disparate as history, philosophy, literary studies, and anthropology had begun to entertain discussion of contextualism and what it held in store for each discipline of knowledge.[1] But while the emphasis on something called contextualism resonated across these disciplines, what was meant by the term, how it was applied (whether to a theory of knowledge, a philosophical system, an anthropological method, or a cultural aesthetic), how to best put contextualism into practice, and what contextualist approaches offered beyond other established humanistic practices varied markedly across these disciplines of knowledge. As even wider audiences began to embrace discussions of contextualism, attempts to define or fix its meaning became more and more difficult, becoming for some an impossibility. Although it appears a difficult task, possibly even antithetical to the stronger versions of the contextualist project, an attempt to convey its varied meanings must be made for readers not familiar with the debates.

This chapter is thus devoted to discussions of the forms of contextualism that have made or are making their way into the study of scientific knowledge. One version of contextualism developed in this discussion has informed the historical problem of the evolutionary synthesis so that

---

[1] This body of literature is vast. For an especially lucid account of the various forms of contextualism in the humanities, see Seymour Chatman, "What Can We Learn from Contextualist Narratology?" *Poetics Today* 11 (1990): 310–28. Similar discussions over contextualism have taken place in architectural theory.

the possibility of a narrative felicitous to the perspective of the historical actors can be constructed. In addition to exposing the theoretical scaffolding for the narrative of *Unifying Biology*, this chapter historically surveys and critically analyzes some of the most recent literature in the history and philosophy of science. It also charts the wider movements in these fields leading to science studies and cultural studies. Because it takes a synchronic, analytical perspective, it departs from the historical (and historicist) style of writing characteristic of other parts of *Unifying Biology*.

SHIFTING GEARS: DEFINING CONTEXTUALISM

The search for some sort of definitional parameters or some kind of boundary around the term "contextual" is made difficult by adherents of especially strong versions of contextualism, who would reject the notion that one can define, fix, limit, or restrict the meaning of literary terms. The same strong contextualists would also reject the notion that one would attempt to essentialize, or to delimit, or to typologize the context (emphasis here on the definite article), especially without the concomitant consideration of the shifting critical position or standpoint of the observer. Nonetheless, for those readers of this text outside these stronger contextualist movements needing "a fix" on some provisional working meaning of the term, it may be helpful to note that the term "culture"—another equally nebulous term[2]—nearly always has something to do with "context." But it should also be noted that the conjunction of the two terms in the often-used phrase "cultural context" would be considered redundant to adherents of strong contextualism; no meaning exists outside of context, and culture represents the processes by which meaning is attached. There is no context outside of culture.

Perhaps one way to understand stronger contextualist movements is to recognize a strong commitment on the part of its practitioners to the view that knowledge is grounded not in foundational, first, or axiomatic principles, but is instead best seen as a cultural artifact (for some it is a product) emerging from more localized and specific contexts of cultural activity. Knowledge is thus culturally embedded and may be seen as an artifact of the culture—a cultural construction—though not necessarily holding any artifice or falseness in its meaning. Rather than upholding

---

[2] In attempting to articulate the meaning of "culture" Stephen Greenblatt states that culture "is a term that is repeatedly used without meaning much of anything at all, a vague gesture toward a dimly perceived ethos." Stephen Greenblatt, "Culture," in Frank Lentricchia and Thomas McLaughlin, eds., *Critical Terms for Literary Study* (Chicago: University of Chicago Press, 1990), p. 225.

the view that knowledge is universal and transcendent, therefore, the movement toward contextualism instead emphasizes the local, situated, and embodied features of knowledge. All knowledge is thus viewed as being historically rooted and culturally embedded; what counts as knowledge is made, fashioned, constructed within a given local context of activity. Because of the localization of knowledge, contextualism in its strong forms also resonates with the multicultural project. Recalling the guiding image of William Blake's *Fall of Man* in chapter 1, the differences between the two are noted by the polarities represented by the center (the unifying element) and the periphery (the dispersive elements) of the watercolor.

THE CONTEXTUAL TURN IN THE HISTORY OF SCIENCE

While the turn toward contextualism has met with opposition and confusion from many of the more traditional approaches within all the relevant fields of knowledge (frequently generating acrimonious debates), it has met the strongest opposition in those fields whose historical development has been to make the greatest commitment to belief in universal, transcendent knowledge.[3] Standard scientific disciplines, the so-called hard sciences on the bottom (or the top, depending on one's vantage point) in the disciplinary ordering of knowledge and now largely compartmentalized and sheltered from currents in the humanities, have been spared these controversies. The softer sciences (some of which are arguably even nonsciences) that border on the humanities, like anthropology, history, and philosophy, have experienced or are presently experiencing the greater turbulent activity as forms of contextualism make their way into general discussion.

The introduction of contextualist theories of knowledge has possibly met with the greatest opposition in the interdisciplinary field of the history and philosophy of science (HPS), where stronger contextualism has only recently been introduced as a viable theory of knowledge.[4] That this opposition has been especially severe is no surprise given the fact that HPS is the field which, in occupying a disciplinary location midway be-

[3] See Paul Forman, "Independence, Not Transcendence, for the Historian of Science," *Isis* 82 (1991): 71–86.

[4] The subject of contextualism in science studies was discussed at a 1990 workshop held at Stanford University. See the forthcoming volume of the proceedings edited by David Stump and Peter Galison entitled *Disunity and Contextualism: Philosophy of Science Studies*, set to appear with Stanford University Press. See also Andrew Pickering, ed., *Science as Practice and Culture* (Chicago: University of Chicago Press, 1992); and idem, *The Mangle of Practice: Time, Agency, and Science* (Chicago: University of Chicago Press, 1995).

tween the sciences and the humanities, serves as a conduit for intellectual exchange between the so-called two cultures. A mixing of approaches from both areas is inevitable, and though frequently generating innovative work, the same mixing has led to some of the most vituperative exchanges in academic circles. The controversies surrounding the application of contextualism to science are so great—and frequently so confusing to the participants—that, taken as a whole, they are possibly the most divisive issues in the history and philosophy of science at the moment.[5]

The intensity of the opposition to contextualism emerges not only from the fact that the interdisciplinarity of the field makes it more possible to mix approaches, but also by the extraordinarily complex range of problems introduced by the application of contextualist philosophy to scientific practice, which relies on the use of instruments, experiments, and modeling procedures, all of which seem to resist the simple discursive analysis common in some forms of contextualism.[6] Given these complications, some of which have been discussed for a range of sciences[7] as well as the range of differences in what can effectively count as science, and to whom, the application of the contextualist project—and what it means for the sciences—thus requires careful examination, especially before it can even begin to help inform historiographic problems like the evolutionary synthesis.

CONTEXTUALIST ACCOUNTS OF SCIENTIFIC KNOWLEDGE: EXTERNALISM, THE SOCIAL CONSTRUCTION OF KNOWLEDGE, AND SCIENCE STUDIES

Despite the severe controversy—and confusion—generated by contextualist approaches in the history and philosophy of science, the number of scholars entertaining discussion and application of some form of contextualist approach has been steadily rising. Reasons for this increase are

[5] These divisions have been brought to the fore, if not aggravated, by Paul R. Gross and Norman Levitt, *Higher Superstition: The Academic Left and Its Quarrels with Science* (Baltimore: Johns Hopkins University Press, 1994). See the special call of alarm in the full-page notice for the National Association of Scholars with the headline: "Science Is Under Attack," *Science* 265 (1994): 1508. The notice continued: "The National Association of Scholars is an organization of academics and independent scholars formed to combat the irrationality and politicization now thriving in university life."

[6] See, for instance, the discussion by Ian Hacking, *Representing and Intervening* (Cambridge: Cambridge University Press, 1983). See also the discussion on some of the practice industry below.

[7] J. E. McGuire and Trevor Melia, "Some Cautionary Strictures on the Writing of the Rhetoric of Science," *Rhetorica* 7 (1989): 87–99.

related to the history of the discipline as a whole. Once a discipline that relied heavily on the historical accounts of scientists-turned historians, the discipline now draws on a generation of scholars who have trained extensively, if not exclusively, as professional historians and philosophers of science.[8] Given this professional transition and the diversity of ways that one can approach both history and science (as well as philosophy), it is understandable that a wide range of historiographic approaches exist, some of which are not just different from the other but also frequently at odds with each other.

At least some (though a lesser part) of the controversy surrounding the use of contextual historiography has to do with the transition to professionalized history and philosophy of science.[9] Having viewed themselves as throwing off the yoke of the scientific disciplines, free to write "objective" accounts of science, and acting as historical critics of science where need be, professional historians and philosophers at best ignore more traditional histories and at worst deride naive accounts of the progression of great "heroes" (nearly all of whom are white males) embarked on an enterprise leading inexorably to the "truth." No longer the mere mnemonic devices, illegitimate children, or "handmaidens" to the scientific disciplines, professional historians of science effectively distanced themselves from their scientific objects of study so much that they frequently lost sight of the same historical objects of study. As noted earlier, even so grand an occasion as the three hundreth anniversary of Newton's *Principia* was hardly worthy of notice by *Isis*, the leading journal of the History of Science Society.

The move for independence was also supported by a long-held prejudice against recent and contemporary science as not being properly historical. (The recent past had too much "noise" in the historical system; only with enough time could one get the proper historical distance or objectivity to make out what had actually happened.) Thus, the professionalized style of historians of science increasingly focused on the "exter-

---

[8] For the historical origin story and the founding father fable for the history of science, see Arnold Thackray and Robert K. Merton, "On Discipline Building: The Paradoxes of George Sarton," *Isis* 63 (1972): 473–95. For data and analysis into stages of the prehistory of the history of science, see Arnold Thackray, "The History of Science Society: Five Phases of Prehistory, Depicted from Diverse Documents," *Isis* 66 (1975): 445–53 among the special celebratory proceedings of the fiftieth anniversary of the History of Science Society included in the same volume. See also idem, "The Pre-History of an Academic Discipline: The Study of the History of Science in the United States, 1891–1941," *Minerva* 18 (1980): 448–73; for another interpretation, see Nathan Reingold, "History of Science Today, 1. Uniformity as Hidden Diversity: History of Science in the United States, 1920–1940," *British Journal for the History of Science* 19 (1986): 243–62. See also n. 3.

[9] See n. 3.

nal" determinants of science (often termed the social, institutional, or generally "cultural" factors), rather than the "internal" determinants of science (the internal logic and structure of the science proper). As well, the same arguments for "distance" from the community of scientists also justified the lack of engagement with scientific or "technical" details which, for recent sciences, involved a major investment of time, training, and disciplining within the scientific culture. It is, after all, much easier and quicker to produce a dissertation, article, or book without worrying unnecessarily about getting "into the scientists' heads" or learning their technical language and practices. Thus, as history of science became professionalized, the professional system, which understandably emphasized rapid productivity, began to selectively empower those scholars most likely *not* to confront scientific culture directly.

The rise in popularity of externalist history of science was also aided and abetted by the rise in social history among wider historical audiences in the 1960s. As an outcome of this movement, historians would no longer contemplate the political and intellectual worlds of elite great men, but would instead look to the everyday life of peasants, workers, and the rank and file to explore how their community structures functioned. Though hardly constituting a unified, monolithic movement, social historians weaned on the antielitist politics of the 1960s effectively swayed historical scholarship so forcefully that intellectual history—the history of the "disembodied" ideas of elite white males—effectively went into demise.[10] Though historians of science have not always followed historiographic trends in wider historical circles, they were sufficiently affected by these larger calls for a more politicized, "socially responsible" history of the commons and increasingly favored the social history of science.

Another reason—not unrelated to the above discussions—for the apparent rise in the popularity of contextualist accounts of science was due to the successful introduction of sociological analyses in the study of science. Although Thomas Kuhn's *Structure of Scientific Revolutions*, which opened discussion of sociological approaches to the study of science,[11] was hardly met with open arms by historians and philosophers of science in the 1960s, sociological analyses of science increasingly began to gain

[10] The introduction of social history did not go unchallenged. See Gertrude Himmelfarb, *The New History and the Old* (Cambridge, Mass.: Belknap, 1987). In recent years the introduction of new cultural history and the application of theoretical perspectives from literary studies have generated controversy. See Gertrude Himmelfarb, *On Looking into the Abyss: Untimely Thoughts on Culture and Society* (New York: Alfred A. Knopf, 1994).

[11] It will be recalled that Kuhn was not a sociologist.

favor through the 1970s and 1980s. This was largely due to the influence of various sociological schools, the most important of which was the Edinburgh school or the "Strong Programme," some of whose adherents also drew and continue to draw on varied anthropological approaches. All these schools offered various social and cultural frameworks for understanding the relationship of science within social structures of production.

Although they had different points of origin, relied on different social theories, and were practiced by vastly differing practitioners, these social/ sociological, external, and "contextual" movements in the history of science began to resonate with each other by the late 1960s. The social construction of knowledge—the view that knowledge was actively fashioned by social determinants and understood through rigorous sociological methods, and that the goal of any such study of science was to deprivilege the "privileged" practice of science—entered serious general discussions, but only with a great deal of controversy.[12]

Though these movements hardly constituted a unified front (some of the more hostile exchanges were between constructivists of various ilks), all bore one thing in common in focusing on the *social* determinants of scientific practice. (Because of their focus on social determinants, they are sometimes also termed externalist, and because of this, they are also termed "contextual" [the emphasis here is on the *social* context]. Though a range of these approaches are classed as contextual [very often in an accusatory tone by unknowing internalists], they are not necessarily allied or even compatible with some of newer, stronger contextual approaches that are making their way in the humanities as part of the larger contextual turn [see the discussion below].) Although these social and sociological approaches had merit in counterbalancing the former internally disembodied histories of elites, they also worked to the detriment of science studies as a whole, effectively serving to reinforce the division between external (and hence social) and internal (and hence intellectual) components of science. Retreating from the scientists' internalistic and privileged location, these social and sociological approaches thus drew dividing lines between the scientific actors and the historical or sociological analysts. Combined with a plethora of other approaches to science coming from feminist circles (of staggering diversity),[13] diverse forms of mul-

---

[12] See Stephen Woolgar for a summary of some of this work. Steve Woolgar, *Science: The Very Idea* (Chichester, Sussex: Ellis Horwood, 1988).

[13] For explication of varied feminist approaches to the study of scientific knowledge, see Joseph Rouse, "Feminism and the Social Construction of Scientific Knowledge," in L. H.

ticultural critiques, and a garden variety of the usual extreme forms of antiscience or science bashing under the guise of criticism, scientific perspectives have been rendered at best obsolete or unreliable, and at worst corrupted; small wonder, then, that at the time of writing a virtual "culture war" has broken out between proponents of science and some socially inclined analysts of scientific study.[14]

One way to temper these debates so that they lead to more constructive exchanges is to tease apart the varied uses of terms like "contextualism" in order to understand changes in meaning across different communities. By far the most able, productive, and compelling introduction of the power of contextualism to inform the history of science and to diminish the distinction between external and internal appeared in 1985. Entitled *Leviathan and the Air-Pump*, this book by Steven Shapin and Simon Schaffer represented the most sophisticated fusing of the move to social history and to the sociology of science—so much so, that it served as a landmark if not a watershed in the history of science for its methodological innovation. Informed by the sociology of knowledge (in Shapin's famous call: "One can either debate the possibility of the sociology of knowledge, or one can get on with the job of doing the thing"),[15] Shapin and Schaffer demonstrated in a historically convincing manner how matters of scientific facts were constructed by the complex interplay of material, social, and literary technology within local contexts of activity in Restoration England. They summarized their argument in a critical statement in their introduction: "We argue that the problem of generating and protecting knowledge is a problem in politics, and conversely, that the problem of political order always involves solutions to the problem of knowledge."[16] So compelling was the historical discussion and the argument for the emergence of early modern experimental science within the sociopolitical context of Restoration England that even historians of science who had long resisted the sociological framework proposed by Thomas Kuhn in *The Structure of Scientific Revolutions* admitted—if somewhat reluctantly—to a view of science as historically rooted and culturally embedded practice. The dichotomy between "internal" and

---

Nelson and J. Wilson, eds., *A Dialogue between Feminism, Science, and the Philosophy of Science* (Dordrecht: D. Reidel, 1996).

[14] See n. 5.

[15] Steven Shapin and Simon Schaffer, *Leviathan and the Air-Pump: Hobbes, Boyle, and the Experimental Life* (Princeton: Princeton University Press, 1985), p. 15. An early version of this phrase appeared in Steven Shapin, "History of Science and Its Sociological Reconstructions," *History of Science* 20 (1982): 157–211.

[16] Shapin and Schaffer, *Leviathan and the Air-Pump*, p. 21.

"external" determinants of scientific practice had been demonstrably collapsed within a "contextualist" scheme of science. At the same time that it demonstrated that the distinction between internal and external could be removed, *Leviathan and the Air-Pump* also served to blur perspectives of the historian, sociologist, and philosopher. The result was what its proponents called "science studies," a transdisciplinary configuration in which history, philosophy, and sociology of science became "inextricably linked."[17]

But the contextualist framework that argued for the social construction of scientific knowledge also introduced an interesting range of problems for future historiographers of science. Emerging, in part, from an allegiance to the sociology of knowledge (and to the Edinburgh "Strong Programme"), the analytic framework in *Leviathan and the Air-Pump* supported a crude form of social constructivism. The end result served to effectively reduce scientific knowledge to sociology. Thus, at the same time that it admitted social and sociological components into scientific practice, it did so at the expense of the intellectual and philosophical features of scientific practice.

Even more problematic in its historiography was the absence of the attempt to understand the perspective of the historical actors. This led effectively to the "silencing" of their voices. In the context of historical movements attempting to problematize and make explicit the relations between historian and historical actor, this feature of *Leviathan and the Air-Pump* grew to become an unacceptable historical method. Methodologically, this problem had grown out of Shapin and Schaffer's sociological writing of history that approached the study of science from an outsider's—or, in their terms, "stranger's"—perspective. The same stranger's perspective was what allowed or gave Shapin and Schaffer their "historical objectivity." But though they could claim to use the tools of sociologists or anthropologists to give such a stranger's or outsider's account of the science of Hobbes and Boyle, they also remained firmly inside the scientific and positivistic sociology of science. Thus, whereas they could argue persuasively for the social construction of scientific knowledge, they were unwilling to apply the same sociology of knowledge to their own practices as historian-analysts. Shapin and Schaffer's argument, and other such attempts to argue for the social construction of scientific knowledge, therefore bore a serious contradiction: while they

[17] See chap. 3 for note of this reconfiguration; and see Chris Raymond, "Scholars Take a New Approach in Studying the Institution of Science," *Chronicle of Higher Education* 9 May 1990, p. A4–A7.

argued against simple-minded scientific empiricism, they argued for similar simple-minded historical empiricism; and at the same time that they acted to "deprivilege" the knowledge-making claims and positions of scientists, they also served to privilege their own knowledge-making claims and positions as historian-analysts of science.[18] For this reason, Shapin and Shaffer's stunning contribution has come into question by recent students of the field of cultural studies (see below). To summarize, *Leviathan and the Air-Pump* was effective in convincing its readers to diminish the distinction between internal and external components of science. In so doing, it opened the door to sociological and anthropological approaches and to a strong version of contextualism, but was ineffective as far as problematizing historiography as *the writing of history*.

MANGLING SCIENCE: THE RISE OF THE "PRACTICE INDUSTRY"

Although *Leviathan and the Air-Pump* generated controversy (as did the introduction of other sociological accounts of science),[19] the book convinced younger historical scholars of the worthiness of sociological and anthropological approaches to the study of science. As workers increasingly entertained transdisciplinary reconfigurations like "science studies," they added to the proliferation of approaches that traveled under the banner of "contextual." Some of these scholars had articulated the notion of practice and linked it with context because they felt that existing studies of science were much too focused on its theoretical aspects. They urged instead a consideration of the actual practice of science. Rather than examine the origin and history of "ideas," which resembled disembodied ghosts, or universalizing theories, which had failed to universalize their claims, and rather than talking about scientists and their science, these workers urged a return to what they termed "context" and "practice."[20] While some returned to close examination of scientists in their laboratories, others went into the field, tried to examine specific institutional sites and the conditions under which disciplinary formation

[18] For this reason, some science studies proponents have turned to the problem of reflexivity. For one attempt to deal with reflexivity, see Malcolm Ashmore, *The Reflexive Thesis: Wrighting Sociology of Scientific Knowledge* (Chicago: University of Chicago Press, 1989).

[19] For a summary of this literature, see Woolgar, *Science*.

[20] For a historically critical account of the rise of "practice" and its relation to science studies and cultural studies, see the excellent chapter entitled "The Significance of Scientific Practices," in Joseph Rouse, *Engaging Science* (Ithaca: Cornell University Press, 1996).

took place, or studied the habits of practice, the accumulation and then transformation of practical skills, the use of instruments, models, and other such procedures. All ultimately shared the goal of understanding how science "actually works." As they sought to return to the laboratory or the site of action to explore how knowledge was locally made, fashioned, and constructed, they all began to simultaneously resonate with sociological and anthropological terms like "practice," which could also serve as a powerful antidote to "theory" and hence a rallying call to those wishing to reform "theory-dominated" history and philosophy of science. While some workers drew on a wide range of theoretical schools of sociology, others began to draw on the cultural theory of anthropologists, and began to explore seriously the tools of sociologists and anthropologists, as well of course, to examine the "tools" of science. As these anthropological approaches turned to ethnographic studies, science itself became a "culture." Because such anthropological analyses also had to examine the discursive or language-based features of cultures, as well as the rites and practices that emerged from and sustained these cultures, they also began to introduce discussion of discourse and practice at the same time that they discussed culture and context.[21] Yet while they all returned to the actions of scientists—as material, social agents of cultures—not all returned to the *perspective* of scientists.

Possibly the most idiosyncratic of these approaches has come from Bruno Latour. In a series of widely read and influential books, Latour has argued for an anthropological approach to understanding science as a culture.[22] This involves "following scientists" around the laboratory and tracing out the process by which facts are made in a laboratory setting. How the individuals arrange themselves within laboratory collectives, how they generate scientific "facts" within the social setting of the scientific community through the use of inscriptional and then persuasive devices involving the marshaling of resources, enrolling of allies, and finally through subsequent "trials of strength" between rival fact-producing cultures, has been mapped systematically by Latour. Latour's work has received much attention and criticism from a range of scholars within sci-

[21] See, for instance, Sharon Traweek, *Beamtimes and Lifetimes: The World of High-Energy Physicists* (Cambridge, Mass.: Harvard University Press, 1988); Pnina Abir-Am, "A Historical Ethnography of a Scientific Anniversary in Molecular Biology," *Social Epistemology* 6 (1992): 323–54.

[22] See Bruno Latour and Steven Woolgar, [1979] *Laboratory Life: The [Social] Construction of Scientific Facts*, 2d (rev.) ed. (Princeton: Princeton University Press, 1987); Bruno Latour, *Science in Action: How to Follow Scientists and Engineers through Society* (Cambridge, Mass.: Harvard University Press, 1986).

ence studies. Among the most unpalatable features of his social framework is his casual overuse of militaristic metaphors, his profoundly
ahistorical orientation, his eager dismissal of cognitive content, and his
rather dim view of human motivators and activities. All these criticisms
stem from his naive acceptance of the colonizing ethnography he has
eagerly adopted. That the power relations of ethnographer to ethnographic object have been problematized, and that ethnographers have
struggled with "empathy" issues to understand the perspective of the
"native" (the same thing as scientists, as Latour would have them) within
a staggering diversity of postcolonial ethnographies that permit the voice
of the "native" to speak (or at the very least to problematize "voices")
seems to have completely escaped Latour (despite the fact that critics like
Donna Haraway have repeatedly pointed this out).[23]

More successful and palatable applications of contextualism and science as "practice" have also made their way to the philosophy of science
through the work of philosophers of science like Ian Hacking, Nancy
Cartwright, and Peter Galison; philosophers like Joseph Rouse; histo-
rian-philosophers like David Stump; historians like Jan Golinski and
Robert Kohler; and sociologists like Adele Clarke, Joan Fujimura, Harry
Collins, Karin Knorr-Cetina, and especially Andrew Pickering.[24] By no
means a unified movement (individuals noted above can be easily sepa-

[23] Discussion of the inclusion/exclusion of "native voices" and appropriate historical and
ethnographic methodology has recently gained significant attention from a wide audience
following Gananath Obeyesekere's revisionist history of the circumstances of the death and
"apotheosis" of Captain Cook. See Gananath Obeyesekere, *The Apotheosis of Captain Cook:
European Mythmaking in the Pacific* (Princeton: Princeton University Press, 1992). This
book attempts to revise Marshall Sahlins' Eurocentric account (according to Obeyesekere).
See Marshall Sahlins, *Islands of History* (Chicago: University of Chicago Press, 1985). For
Sahlins' response to his critic, see Marshall Sahlins, *How "Natives" Think, About Captain
Cook, for Example* (Chicago: University of Chicago Press, 1995). For an excellent review
that discusses both positions, see Clifford Geertz, "Culture War," *The New York Review of
Books*, 30 November 1995, pp. 4–6. For a discussion of the knowledge–power nexus in
science studies, see Joseph Rouse, *Knowledge and Power: Toward a Political Philosophy of
Science* (Ithaca: Cornell University Press, 1987).

[24] See nn. 4 and 23. See also Timothy Lenoir and Yehuda Elkana, "Practice, Context and
the Dialogue between Theory and Experiment," *Science in Context* 2 (1988); and Jan
Golinski, "The Theory of Practice and the Practice of Theory: Sociological Approaches in
the History of Science," *Isis* 81 (1990): 492–505; Adele E. Clarke and Joan H. Fujimura,
The Right Tools for the Job: At Work in Twentieth-Century Life Sciences (Princeton: Princeton
University Press, 1992); Harry Collins, *Changing Order: Replication and Induction in Scientific Practice*, 2d ed. (Chicago: University of Chicago Press, 1992); Robert Kohler, *Lords of
the Fly: Drosophila Genetics and the Experimental Life* (Chicago: University of Chicago
Press, 1994); Karin Knorr-Cetina, *The Manufacture of Knowledge: An Essay on the Construc-
tivist and Contextual Nature of Science* (Oxford: Pergamon, 1981); and idem, "The Couch,
the Cathedral and the Laboratory: On the Relationship between Experiment and Laboratory in Science," in Andrew Pickering, ed., *Science as Practice and Culture* (Chicago: University of Chicago Press, 1992), pp. 113–38.

rated on other nonpractical grounds), all primarily have envisioned science as practice and mostly examined sciences like physics or physicalistic life sciences.[25]

All also share a dissatisfaction with the merely theoretical (or what I would also call the ideational or intellectual components of science) and either eliminate the latter or reduce it to material knowledge. As part of this "move away from theory-dominated" and "representational" accounts of science, the focus instead is on the practice of science to understand what role instruments, models, experiments, and other such "interventionist" procedures play in science. Although their retreat from theory and representational practices permits the growing "practice industry" to successfully avoid charges of "destructive relativism" in favor of some sort of instrumentalist/realist, rationalist, and pragmatist theories of knowledge (and subtle variations on these themes), their retreat from theory into practice has only limited validity to those sciences whose procedures effectively diminish theory, representational practices, or narrative practices—sciences like experimental physics. What sort of "contextualist" account could the "practice industry" then give of historical sciences like archaeology, cosmology, geology, and evolutionary biology, whose textuality and narrativity is the most transparent feature of the science and which have limited use of material, or observational, or instrumentally determined evidence? And where does the critical problem of historical interpretation come in? Equally problematic with this practice-oriented philosophy is an obsession with the same underlying, defining question of "how science *actually* works." For historians this is a problematic question: that "science" has a historicity (and temporality) that may defy attempts to define, essentialize, or typologize "it," that it may, like a "form of life," defy attempts to freeze it (synchronically) for analysis, and that equal consideration should be given to historical questions such as how science came to be have received too little discussion. As Andrew Pickering—the leading pundit of practice—has noted in his latest reflections, the notion of time and the fact that science emerges within a set of temporally embedded practices is indeed an important consideration for understanding science.[26]

---

[25] For the most recent account of science as practice in physics, see Brian Baigrie, "Scientific Practice: The View from the Tabletop," in Jed Buchwald, ed., *Scientific Practice: Theories and Stories of Physics* (Chicago: University of Chicago Press, 1995), pp. 87–122. See also Peter Galison's contribution to the volume entitled "Context and Constraints," pp. 13–41.

[26] Andrew Pickering, *The Mangle of Practice: Time, Agency, and Science* (Chicago: University of Chicago Press, 1995). Sharon Traweek has explored scientists' constructions of time,

Most problematic of all is that the emphasis on theory tilts understanding of science too heavily in the direction of its paired duality (as in theory/practice). Hence, such approaches reinforce practice to the exclusion or diminishment of theory. That science consists of more than just material practices, that it functions as a belief system whose narratives lend coherence and meaning to the community (or any of the myriad terms that have been invented, like life-world, cosmology, weltanschauung, forms of life, thought-collective, paradigm, and discursive mentalité, etc.) seems to have gained little serious discussion (despite the long-held importance of such approaches). Worse still, to define science as the "doing of work," as Clarke and Fujimura (among others) have attempted to do, or to define science as a cultural "product" or "good," to be sold, purchased, possessed, or consumed, runs into the danger of extinguishing the existential dimensions of science (no doubt these existential concerns are most apparent in sciences like evolution). These "workers" (their own choice of terminology) seem to have forgotten the commonsense philosophical wisdom that "we do not have to do, to be"; and our meaningful existence is not solely determined by work, products, or other such artifacts of existence. That science emerges from, and is inextricably linked to, other, more humanistic, cultural concerns that serve existential and aesthetic needs, and that it may be modeled after aesthetics in the way of being an expression of humanistic desires; that it may be part of a culturally embedded belief system from which groups have derived values, seems to have been forgotten completely in the stampede to retreat from theory. To sum: the problem with much of the practice-oriented contextual philosophy is an overemphasis on the material culture of science and the materiality of knowledge, which accompanies the diminishment, elimination, or exclusion of the narrative worlds, discursive mentalities, or—in whatever word we choose here—the *perspective* of the scientists.

SCIENCE AS DISCOURSE AND CULTURE: POWER/KNOWLEDGE IN THE CULTURAL HISTORY/CULTURAL STUDY OF SCIENTIFIC KNOWLEDGE AND THE ANTHROPOLOGY OF KNOWLEDGE

Yet other approaches have come from movements within historical circles, especially from historians of the early modern period and historians

---

though she has not explicitly noted that temporality could be applied to narrative pattern; see Traweek, *Beamtimes and Lifetimes.*

of the eighteenth century. Most of this work is part of, or feeds into, movements in wider historical circles associated with cultural history or cultural studies.[27] Mario Biagioli, Paula Findlen, Steve Shapin, and Jay Tribby have actively reworked contextualist historiography of science by exploring the emergence of science within court culture, for instance.[28] Biagioli's theoretical grounding for his contextual theory of knowledge is the most lucid and complete application of cultural theory to the history of science to date. But to deal with the historicity of scientific knowledge—as he so wishes—Biagioli would have had to explore the narrativity of scientific knowledge in order to rework the grand narrative of the history of science that constructs, locates, and determines the character of Galileo.[29] Without this consideration, his Galileo becomes hardly more than a courtly parvenu: his passionate aestheticism is forgotten, and the language of his physics is silenced.

The importance of narratives and the fundamental narrativity of all knowledge underscore the work of both Misia Landau and Donna Haraway.[30] Both examine the narratives of anthropology, clearly a science rich in use of narrative. But while both share a common vision of reworking scientific narratives into narratives of emancipation, both do not succeed equally well in their project.

For Landau, anthropological narratives follow the narrative rules or patterns that undergird much of human thought. In other words, these narratives are versions of "archetypes" (a notion borrowed from Northrop Frye applied to folktale analysis by Vladimir Propp) or narratives

[27] By far the best introduction to the cultural study of scientific knowledge is Joseph Rouse, "What Are Cultural Studies of Scientific Knowledge?" *Configurations* 1 (1992): 1–22. For an introduction to the general area, see L. Grossberg, C. Nelson, and P. Treichler, eds., *Cultural Studies* (New York: Routledge, 1992), and see Fred Inglis, "Cultural Studies," *Times Literary Supplement*, 27 May 1994. As noted in chap. 3, the introduction of cultural studies has met with some resistance from sociologists associated with the sociological study of scientific knowledge. An example of some of this resistance can be found in H. M. Collins, "Review of *We Have Never Been Modern*, by Bruno Latour," in *Isis* 85 (1994): 672–74; and Peter Dear, "Cultural History of Science: An Overview with Reflections," *Science, Technology, and Human Values* 20 (1995): 150–70.

[28] Mario Biagioli, *Galileo, Courtier: The Practice of Science in the Culture of Absolutism* (Chicago: University of Chicago Press, 1993); Paula Findlen, "Jokes of Nature and Jokes of Knowledge: The Playfulness of Scientific Discourse in Early Modern Europe," *Renaissance Quarterly* 43 (1990): 292–331; Steven Shapin, *A Social History of Truth: Civility and Science in Seventeenth-Century England* (Chicago: University of Chicago Press, 1994); Jay Tribby, "Cooking (with) Clio and Cleo: Eloquence and Experiment in Seventeenth Century Florence," *Journal of the History of Ideas* 52 (1991): 417–39; idem, "Club Medici: Natural Experiment and the Imagineering of Tuscany," *Configurations* 2 (1994): 215–35.

[29] For more explication, see H. Aram Veeser, ed., *The New Historicism* (New York: Routledge, Chapman and Hall, 1989).

[30] Misia Landau, *Narratives of Human Evolution* (New Haven: Yale University Press, 1991); Donna Haraway, *Primate Visions: Gender, Race, and Nature in the World of Modern Science* (New York: Routledge, 1989).

that obey a priori rules that have biological points of origin. Liberation, for Landau, comes from reworking such narratives; yet her choice of theory effectively subverts her project. If all narratives follow a priori rules, especially governed by biology, then how can one rewrite biological narratives (which, by definition, are a priori)?[31]

For Haraway, the structural form and function of the narrative is not so much the concern (she would decry the use of such structuralist/functional theories, in fact); instead, the stories told of science by scientists are of interest in and of themselves for what they reveal of undergirding values that emerge from, and sustain, power relations in Western culture. Drawing creatively on some of the most recent postcolonial ethnography, Haraway's historical practice seeks to deconstruct, disrupt, or diffuse existing power structures inherent in all social systems, through which race, class, and gender become structured. In keeping with postcolonial ethnography, "objectivity" becomes a function of the observer's critical positionality (his or her vantage point or point of view). Positioning herself as feminist critic, Haraway appropriates critical tools that enable the reworking of the narratives of science that serve to unmask, expose, and disrupt notions of race, class, and gender embedded within the scientific system of power relations. Critical knowledge, in her view, becomes a tool for social action.[32] Haraway's work is especially noteworthy for the historian of evolution as her subject matter is evolutionary theory, and more specifically primate biology in the period up to, including, and after the evolutionary synthesis. As she points out in a startling, and terrifying manner, stories of evolution (like all stories about ourselves) cannot but be vehicles for the transmission of hidden values pertaining to race, class, and gender.[33]

Haraway's use of narrative theory for science makes for an able inspiration to the present project, with one notable revision. Undergirding Haraway's narratives of emancipation and cultural theory is a largely static structure (or structurated, to use the British term) or view of culture. Though Haraway would deny such a strictly static view, she would agree that knowledge as a tool for social action becomes an enabling device to

---

[31] For more detailed explication and criticism, see V. B. Smocovitis, "Review of Misia Landau, *Narratives of Human Evolution,*" *Journal of the History of Biology* 26 (1993): 147–63.

[32] See also her corpus of work, including her essays, in Donna Haraway, *Simians, Cyborgs, and Women: The Reinvention of Nature* (London: Free Association Books, 1991); and idem, "The Promises of Monsters: A Regenerative Politics for Inappropriate/d Others," in L. Grossberg, C. Nelson, and P. Treichler, eds., *Cultural Studies* (New York: Routledge, 1992), pp. 295–337.

[33] See n. 30.

break those same structures, in order to give agency and empowerment to the subordinated "others." Here a change of view in sociocultural theory inspired by evolutionists Niles Eldredge and Stephen Jay Gould (lending weight to the linkage between evolutionary theory and sociocultural theory) may critically shift Haraway's position and line of question. Supposing that cultures (like species) are always in a state of flux, the anomalies, in this view, are not the periods of change but the periods of stasis. Reframing the question thus: Why should students of culture seek to explain moments of historical or cultural change, or seek a theory of knowledge that will "bring forth change"? Would it not be more revelatory and fruitful—and ultimately more liberating—to explain moments of lack of change, moments of cultural stasis, or moments that serve to stricture or constrain history and culture and science?[34]

Returning to the problem of the synthesis and the problem of such historical "events," can we not also view the rise of a historical event—and a mutually consented historical "event"—as such a moment of stricture? One that, among other things, alters and remolds and then structures the identities of its members? In fact, one way that we can understand the number of commentators on the subject of the evolutionary synthesis is to view it as such a "defining moment" in the collective history and memory of evolutionary biology—so much so that the evolutionary synthesis has constructed the identities of all evolutionary biologists (not unlike the defining moment of the "French Revolution" for the nation-state called France; I return to this in the discussion below).

SITUATING TEXT IN CONTEXT:
THE RETURN OF INTELLECTUAL HISTORY
AND THE VOICE OF THE SCIENTIST

"Theoretical" concerns aside, how exactly does one proceed to write history within a contextualist framework that addresses all the above problems? How can contextualist approaches help inform historical work on the most practically immediate levels? And how can such contextualism help illuminate the historical problem of the evolutionary synthesis?

From yet another direction, another form of contextualism had been making its way to historical audiences, especially in the late 1980s and early 1990s. As part of a profound rethinking of intellectual history, varied movements began to view contextualism in its literary terms.[35] Al-

---

[34] This bears some resemblance to the cultual theory(ies) of Michel Foucault.
[35] Dominick La Capra and Steve Kaplan, eds., *Rethinking Intellectual History: Text, Con-*

though some of these discussions took place on theoretical playing fields,[36] a small number of intellectual historians began to put to fruitful use a vision of context in its literal meaning of "in or within the text." Most ably applied by intellectual historians like Keith Michael Baker and Carlo Ginsburg, French cultural historians like François Furet, and historians associated with the new cultural history like Roger Chartier and Lynn Hunt, a reconfiguration of literary and historical practice took place that effectively removed the split between social and intellectual histories, within a history of what they termed "discourse."[37] Within this view, historians could effectively collapse approaches to history from the bottom up (social history) and from the top down (intellectual) within a history of "discourse" that operated in the "middle." Resonating with movements in cultural anthropology and cultural theory, historical knowledge could be viewed as (con)textual, in accord with forms of conversation, or in some cases in the forms of dialogues between texts. In this view, what had formerly been termed the history of ideas or intellectual history was reinvented as the history of discourse or new cultural history, but with no disembodied ideas, few or no "originary points," and little or no distinction between popular and elite and other such cultural divisions (see also the discussion below).

Following this theoretical scaffolding, contextual historical accounts seek to situate text within text, stressing the polysemous nature of any reading. Methodologically, the stress is on the close reading and reproduction of texts, which is another way of emphasizing the interpretive nature of historical practice and at the same time giving "voice" to the

---

texts, and Language (Ithaca: Cornell University Press, 1983). See also Donald R. Kelley, "What Is Happening to the History of Ideas?" *Journal of the History of Ideas* 51 (1990): 3–25.

[36] See Dominick La Capra's *History and Criticism* (Ithaca: Cornell University Press, 1985).

[37] Keith Michael Baker, *Inventing the French Revolution* (Cambridge: Cambridge University Press, 1990); Carlo Ginsburg, *The Cheese and the Worms: The Cosmos of a Sixteenth Century Miller*, trans. John and Anne Tedeschi (New York: Penguin, 1982; orig. pub. as *Il formaggio e i vermi: Il cosmo di un mugnaio del '500* [Giulio Einaudi Editore, 1976]; and first trans. Johns Hopkins University Press, 1980); François Furet, *Interpreting the French Revolution*, trans. Elborg Forster (Cambridge: Cambridge University Press, 1981; orig. pub. as *Penser la Révolution Française* [Editions Gallimard, 1978]); Roger Chartier, *Cultural History: Between Practices and Representations*, trans. Lydia G. Cochrane (Ithaca: Cornell University Press, 1988); Lynn Hunt, ed., *The New Cultural History* (Berkeley: University of California Press, 1989). See also the earlier discussions on the history of mentalities: Patrick H. Hutton, "The History of Mentalities: The New Map of Cultural History," *History and Theory* 20 (1981): 237–59; and see also idem, *History as an Art of Memory* (Burlington: University of Vermont, 1993). See also the collection of essays in Michel Vovelle, *Ideologies and Mentalities*, trans. Eamon O'Flaherty (Chicago: University of Chicago Press, 1990).

text. Combined with forms of cultural or ethnographic theory, which recognizes the critical positionality of the historian and the system of power relations between historian and historical actor, the history of discourse adapted to the history of science is also a way of returning and recovering the perspective of the historical actor.[38] Here it should be noted that such histories (like ethnographies) that silence the perspective of the historical actor are not only historiographically, but also morally, politically, and epistemically bankrupt. The voice of the narrator, it should be noted, comes through the text proper as it no longer presents itself as a "stranger's account." It is, instead, an account within culture (for all readers and writers are "inside" some cultural framework) that adopts critical tools (mostly linguistic) that disrupt conventions, rituals, and practices so that a "defamiliarization" takes place;[39] the hopeful outcome is narrators who can situate their own being or voice within the historical narrative, effectively *writing themselves into the story*. For this reason, some postcolonial ethnographers intentionally play with reflexive modes of inquiry.[40]

An additional feature highlighted in historiographic models that play on narrativity and postcolonial ethnography is an emphasis on the script-like nature of the narrative pattern. Whether the script be for the unfolding of a megaevent like the French Revolution[41] or for a microevent like the unfolding of a life, the narrative plays itself out through this script or "runs" itself in the historical actors and their historical accounts. Similarly, knowledge of science emerges from the writing of grand historical narratives that function like scripts that run themselves from scientific

[38] See the reconstruction of the worldview of Menocchio, a sixteenth-century miller in Ginsburg, *The Cheese and the Worms*.

[39] Robert Darnton has played with similar anthropological tools to familiarize himself with historically "other" cultures. See his essay on the "Great Cat Massacre" for a case in which Darnton uses an examination of a "strange" incident (the cat massacre) as a way of gaining access to a past and "foreign" culture. My position here flips Darnton's use of familiarization into defamiliarization, for I would argue that all viewers are inside some cultural framework; hence the importance of critical positionality. See Robert Darnton, *The Great Cat Massacre and Other Episodes in French Cultural History* (New York: Basic, 1984). For a history and discussion of historical objectivity, see Peter Novick, *That Noble Dream: The "Objectivity Question" and the American Historical Profession* (Cambridge: Cambridge University Press, 1988).

[40] Donna Haraway's work deliberately plays with self-reflexive forms. For another example of self-reflexivity in ethnographic practice, and for some sources into postcolonial ethnography, see Renato Resaldo, *Culture and Truth: The Remaking of Social Analysis* (Boston: Beacon, 1989). Both take self-reflexivity as a given.

[41] See François Furet's treatment of the French Revolution: François Furet, *Interpreting the French Revolution*, trans. Elborg Foster (New York: Cambridge University Press, 1981; orig. pub. as *Penser la Révolution Française* [Editions Gallimard, 1978]).

microevents to megaevents and that ultimately lend coherence to the scientific project and lead to the construction and delineation of a collective memory.

GROUNDING DISCUSSION: A HISTORICAL EXAMPLE FROM DARWIN STUDIES

Within the history of evolution, one central concern has been the rise of Darwinism and Darwin's "evolutionary theory" in the nineteenth century. Historians of ideas have conventionally viewed Darwin as a revolutionary thinker who introduced a dynamic view of organic change. Questions in Darwin studies have traditionally taken the following direction: Was Darwin a product of his age? Was evolutionary theory "in the "air," especially given the simultaneous codiscovery by Alfred Russel Wallace? What was it in Darwin's theory that was so revolutionary or original if others were independently deriving similar theories of organic change? A contextualist here would rework narratives conventionally disengaged from the narrative of evolution, like the wider narrative of the history of the "West" and narrower personal narrative of the figure of Darwin, in order to bring many such narratives together. If one assumes that such engagements between narratives always exist, rather than that they are disengaged (so that one must then demonstrate engagements), then it is possible to reframe the questions posed. In more familiar terms, contextualizing involves bringing previously disengaged narratives together in an overlapping mode within a rewoven grander narrative. The problem in this contextually polysemous scheme is not to account for "connections" or "causal influences" but to account for the dislocations or breaks between the narratives. A contextualist thus begins his or her historical work assuming that such "connections" exist within a larger discursive formation, and may begin to reweave another story, possibly with a view of explaining the breaks or dislocations within such a discursive formation.

Returning to the concrete historical example from evolutionary studies, the question becomes not whether Darwin was a "product of his age" but instead what made us think that he was disengaged from his age. So, too, in this contextualist history Darwin is hardly an "original" thinker for there are few "originary" points in the history of discourse, but is instead himself a part of, or a "node," within a discursive network or formation that actively constructed *him*. In this view Darwin was not revolutionary but conservative, and all such scientific "revolutions" become conservative moments of ordering the world. That species "transmute" and "transform" had been part of pre-*Origin* scientific discourse. Darwin himself introduced his own "descent with modification" that became "evolution" and which made possible a remarkably orderly view of organic change, given the alternatives. Within this history, if Darwin had not reordered the world, someone else would have, for the script for evolution was part of the longer and grander script of the "Enlightened West." In this sense, Darwin's historical "other" was his codiscoverer, Alfred Russel Wallace. But as we know from the story, especially as told by Janet Browne, Wallace and Darwin were not exactly located equivocally in the Victorian scientific—or economic—milieu.[42]

The success of contextualist historiography in the example noted above clearly depends on the existence of narratives that can be rethought, revised, and then rewritten. Thus, this form of contextualism accompanies a historical discipline that has reached some level of maturity (in a sense, the texts and narratives must have accumulated); but similar contextualist methods can be used for less mature disciplines or historical subjects. To sum up: this version of contextualism upholds the belief that knowledge is contextual (in and within text); it emphasizes the close reading of texts with attention to precise use of language, and stresses both the polysemy in interpretive readings and the interwoven nature of narratives. The goal of this contextualist project is to narrate an account that allows the voices and perspectives of the historical actors to speak along with other historical voices and the narrator(s), all of whom have written themselves into the story.

CONTEXTUALIZING THE EVOLUTIONARY SYNTHESIS

Like the defining moment of the French Revolution in Furet's analysis, the evolutionary synthesis has formed a similar moment for constructing the memories and identities of members of the discipline of evolutionary biology (so, too, Charles Darwin becomes the "founding father" of the story). As insider to his story (Furet is a French historian writing on the defining moment in the construction of the cultural identity of France, and therefore his own identity), Furet has framed the interpretive problem of the Revolution in the following way: "The Rev-

[42] Janet Browne, *Voyaging: Volume One of a Biography of Charles Darwin* (New York: Alfred A. Knopf, 1995).

olution does not simply 'explain' our contemporary history; it *is* our contemporary history."[43]

Given this infinite regress of mirror-like perspectives, how then is the dream of historical objectivity to be fulfilled? Furet here writes: "There is no such thing as 'innocent' historical interpretation, and written history is itself located in history, indeed *is* history, the product of an inherently unstable relationship between the present and the past, a merging of the particular mind with the vast field of its potential topics of study in the past."[44]

Given Furet's admonitions, how can we adjust our critical position so that we can tilt our mirror-like "insider's view"? One way to gain critical distance is to view an object from a distant point of view or, in Barbara Tuchman's terms, "a distant mirror." Sciences encased in the positivist ordering of knowledge can be viewed from postpositivist or more literary perspectives that overturn the ordering that grounds the humanities and social sciences in the sciences. From such an inverted position, the methods of the humanities are brought to bear on understanding the sciences. Transferring from literary to visual metaphors, we may use optical devices, like low-magnification lenses (alternating when possible with high-magnification lenses) to give form to the historical object. Within this synoptic view objects—or narratives far apart—may be brought together; diverging points of view may be "unified" within a syncretic account.

We may thus begin to view the historical problem of the synthesis as an interwoven series or plexus of narratives disengaged from other historical narratives. *Unifying Biology* is thus an attempt to rework narratives from political, social, philosophical, personal, disciplinary, and intellectual realms within the narrative of the "Enlightened West." Contextualizing means bringing together such narratives to weave a coherent story. Attention is paid, as much as possible, to the terms and their filiations used by the scientific authors/historical actors.

[43] Furet, *Interpreting the French Revolution*, p. 3.
[44] Ibid., p. 1.

PART TWO

The Narrative of *Unifying Biology*

CHAPTER FIVE

The Narrative of *Unifying Biology*: The Evolutionary Synthesis and Evolutionary Biology

> We didn't sit down and forge a synthesis. We all knew each other's writings; all spoke with each other. We all had the same goal, which was simply to understand fully the evolutionary process. . . . By combining our knowledge, we managed to straighten out all the conflicts and disagreements so that finally a united picture of evolution emerged.
>
> Ernst Mayr, *The Omni Interviews*

> The theory of evolution is quite rightly called the greatest unifying theory in biology.
>
> Ernst Mayr, *Populations, Species and Evolution*

THE STRUGGLE to unify the biological sciences is one of the central features of the history of biology. Emerging only in the nineteenth century,[1] biology was characterized by disunity to such an extent and for so long that repeated attempts to unify this science through professional societies proved to be a nearly impossible task. Charting the rocky road toward organized biology in America during the 1889–1923 period—a key

---

[1] Locating the emergence of biology as a legitimate and autonomous science has been a contentious issue for historians of biology. Though the term was coined in the early years of the nineteenth century, an autonomous science of life was not as strongly defensible until evolution was articulated. Only with evolution, which defied reduction to physics and chemistry because of "metaphysical" components (these include arguments for emergentism, teleology, historicism, and those included herein) at the same time that it introduced a causo-mechanical agent for evolutionary change, could biology claim autonomy and legitimacy. The beginnings of this process took place in Thomas Henry Huxley's England, and most likely in the thought of Huxley himself, who adopted the term "evolution" and thereby linked Darwin's theory of descent (with modification) to embryology, physiology, and biology. Huxley had the following to say on the emergence of biology as a science: "the conscious attempt to construct a complete science of Biology hardly dates further back than Treviranus and Lamarck, at the beginning of this century, while it has received its strongest impulse, in our day, from Darwin" (*The Crayfish: An Introduction to the Study of Zoology* [London: C. Kegan, Paul and Trench, 1884], p. 4). Huxley may be viewed as a chief discipline builder for biology. My argument is supported by the recent work of Joseph

period for the institutionalization of biology—historian Toby Appel concluded: "Numerous biological sciences were established in America, but no unified science of biology."[2] So formidable was this task that the hope of ever formulating a unified biological society representing a unified science of biology appeared to have been largely abandoned by 1923.

By the early 1950s, however, the organization of biological knowledge had been greatly transformed. With the formation of the American Institute of Biological Sciences, the first umbrella-like organization representing the heterogeneous practices of the biological sciences was intact. So, too, departments of zoology and botany had begun to give way to the more synthetic rubric of "biology." Simultaneously, there had also grown an awareness of a feeling of unity within the biological sciences.[3] So powerful was the conviction that biology had become a unified science that G. G. Simpson could introduce his biology textbook of 1957 with the following assertion:

> This book is based on strong convictions. We believe that there is a unified science of life, a general biology that is distinct from a shotgun marriage of botany and zoology, or any others of the special life sciences. We believe that this science has a body of established and work-

Caron, "'Biology' in the Life Sciences: A Historiographical Contribution," *History of Science* 26 (1988): 223–68; see also Gerald Geison, *Michael Foster and the Cambridge School of Physiology* (Princeton: Princeton University Press, 1978). For a good example of the origin of biology and its location and history in relation to physics and chemistry in the context of positivist theory in the nineteenth century, see H. W. Conn, *The Story of Life's Mechanism* (London: George Newnes, 1899). He writes: "In recent years biology has been spoken of as a new science. Thirty years ago departments of biology were practically unknown in educational institutions. To-day none of our higher institutions of learning considers itself equipped without such a department" (p. 9). See also his subsequent history of the new science of biology that followed the trajectory of physics and chemistry, dispensed with mysticism, and favored explanatory laws. One could make a strong argument that Ernst Haeckel was as instrumental to biological discipline building in the German national context. For another interesting discussion of the origins of biology in the context of German physiology, see Timothy Lenoir, *The Strategy of Life* (Dordrecht: D. Reidel, 1982; repr. University of Chicago Press, 1989).

[2] Toby Appel, "Organizing Biology: The American Society of Naturalists and Its 'Affiliated Societies,' 1883–1923," in Ronald Rainger, Keith R. Benson, and Jane Maienschein, eds., *The American Development of Biology* (Philadelphia: University of Pennsylvania Press, 1988), pp. 87–120.

[3] Hamilton Cravens has pointed out that the 1920–50 period also witnessed movements to support interdisciplinary scholarship in America. During this period institutional and intellectual networks were assembled to lend an increasing feeling of unity. See Hamilton Cravens, "Behaviorism Revisited: Developmental Science, the Maturation Theory, and the Biological Basis of the Human Mind, 1920s-1950s," in Keith R. Benson, Jane Maienschein, and Ronald Rainger, eds., *The Expansion of American Biology* (New Brunswick, N.J.: Rutgers University Press, 1991), pp. 133–63.

ing principles. We believe that literally nothing on earth is more important to a rational living being than basic acquaintance with those principles.[4]

The years between 1923 and 1950, which spanned the interwar period and World War II, were tumultuous years rocked by global events and global disturbances. The culmination of the "modern period," these years witnessed the rise of international artistic and political movements, as well as related intellectual and philosophical movements.[5] Within the philosophical milieu, members of the Vienna Circle—the logical heirs to positivist philosophy—were at this time actively embarked on an Enlightenment ideal to unify all the sciences within a coherent worldview.[6] Within the biological sciences themselves, this interval of time witnessed a major disciplinary realignment as evolutionary biology emerged as a discipline of the biological sciences. Most important for the unification of biology, these were also the years long-recognized by historians of biology as constituting the period of the evolutionary synthesis.[7]

The present story weaves these threads together into a historical "grand narrative" leading to the belief that the biological sciences had become unified. Close examination of the process of unifying biology also sheds light on a central problem of the history of biology—namely, the evolutionary synthesis. Three continuous phases will emerge pertaining to the synthesis: the unification of biology through the articulation of Theodosius Dobzhansky's population-oriented, evolutionary genetics framework (derived from the work of systematist-naturalists in combination with classical and theoretical population geneticists), the consequent binding of the heterogeneous practices of biology (through a series of dialogues between workers), and the publication of Julian Huxley's *Evolution: The Modern Synthesis* with its articulation of a liberal, humanistic,

---

[4] This excerpt is taken from the preface to the first edition (1957) of George Gaylord Simpson, Colin S. Pittendrigh, and Lewis Tiffany, *Life: An Introduction to Biology*; reprinted in George Gaylord Simpson and William S. Beck, *Life: An Introduction to Biology*, 2d ed. (New York: Harcourt, Brace, and World, 1965), p. v.

[5] Peter Galison has brought into relief the interplay of these artistic, political, and philosophical movements: see "Aufbau/Bauhaus: Logical Positivism and Architectural Modernism," *Critical Inquiry* 16 (1990): 709–52; and see n. 6.

[6] For an article assessing the impact of Machian positivism and the Unity of Science Movement in the United States especially, see Gerald Holton, "Ernst Mach and the Fortunes of Positivism in America," *Isis* 83 (1992): 27–60. See the discussion below on the positivistic goal of unification through reduction.

[7] By far the most comprehensive treatment of the synthesis is Ernst Mayr and William B. Provine, eds., *The Evolutionary Synthesis* (Cambridge, Mass.: Harvard University Press, 1980).

and secular worldview.[8] The evolutionary synthesis was part of the process of unifying biology. Evolution, purged of unacceptable metaphysical elements, became the "central science" of biology that bound together, and grounded, the heterogeneous practices of biology into a unified and progressive science. The science of evolutionary biology—reworked into an experimental and quantitative practice—also emerged at this time. Biology itself in turn became a unified science to rival Newtonian physics and chemistry, but in a manner that also preserved its autonomous status. The unified biology—through the central science of evolution as it emerged from the synthesis—would in turn be centrally situated within the positivist ordering of knowledge. The "architects" of the evolutionary synthesis would henceforth function as the "unifiers," preserving the whole of the disciplinary ordering of knowledge and an Enlightened worldview.

WOODGER'S *Biological Principles*, THE UNITY OF SCIENCE, AND THE INDEPENDENCE OF BIOLOGICAL SCIENCE

The abandonment of the hope that a unified science of biology could ever be possible took place at a time when the foundations and status of biology were being questioned. Within an increasingly positivist philosophical framework, biology, with its remnants of vitalistic thinking and nonrigorous methodology, was seen as filled with speculation. One outstanding critic—but hopeful reformer—of biology was J. H. Woodger, who in 1929 published his *Biological Principles*.[9] In this book, Woodger hurled criticism after criticism at what he viewed as a science in its infancy and rife with metaphysics. Still in the "metaphysical stage" of its development, biology would come of age only after critical thought had purged it of its more speculative features and after fundamental axiomatic principles had been established. Only after biology had paid "critical attention to the purification of its concepts," and only by "making more

---

[8] Julian Huxley, *Evolution: The Modern Synthesis* (London: Allen and Unwin, 1942).

[9] For biographical information on Woodger, see the "curriculum vitae" by W. F. Floyd and F. T. C. Harris included in the special volume of essays collected in honor of his seventieth birthday: John R. Gregg and F. T. C. Harris, eds., *Form and Strategy in Science: Studies Dedicated to Joseph Henry Woodger on the Occasion of His Seventieth Birthday* (Dordrecht: D. Reidel, 1964). Woodger began his career in zoology but turned to the philosophy of biology after a 1926 visit to Vienna. His subsequent turn to philosophy culminated with the publication of his *Biological Principles* in 1929, for which he received his D.Sc. degree.

sure of its foundations," would it become a mature science.[10] For Wood-ger, biologists, who thought they had found their Newton in Darwin, were mistaken, since biology had not yet reached a stage in its development comparable to eighteenth-century physics. Meant to imitate Robert Boyle's "Sceptical Chemist," Woodger's book would function as the "Sceptical Biologist."

> If we make a general survey of biological science we find that it suffers from cleavages of a kind and to a degree which is unknown in such a well unified science as, for example, chemistry. Long ago it has under-gone that inevitable process of subdivision into special branches which we find in other sciences, but in biology this has been accompanied by a characteristic divergence of method and outlook between the expo-nents of the several branches which has tended to exaggerate their dif-ferences and has even led to certain traditional feuds between them. This process of fragmentation continues, and with it increases the time and labour requisite for obtaining a proper acquaintance with any par-ticular branch.[11]

To Woodger, this fragmentation was increasingly alarming, for it seemed that the biological sciences could never rival the unity of more mature sciences like physics and chemistry. This did not bode well for the establishment of a unified science. One way to achieve such unity in the face of fragmentation would be through the articulation of a central bio-logical principle of great "unifying power," a generalization that would "knit the several branches of biology." Woodger wrote: "But whereas in some sciences this process has been accompanied by the attainment of generalizations which have tended to knit the several branches into a single whole, in biology the disruptive process has not been compensated by the help of any principle of such unifying power, and the possibility of a unified biology seems to recede more and more from our grasp."[12] Repeated attempts to unify and synthesize the diverging branches of bi-ology had only brought out underlying differences, however, and had made biology's unstable grounding all the more evident. Woodger wrote:

> The general theoretical results which have been reached by investiga-tion along the lines of physiology, experimental morphology, genetics,

[10] J. H. Woodger, *Biological Principles: A Critical Study* (London: Routledge and Kegan Paul, 1929), p. 84.
[11] Ibid., p. 11.
[12] Ibid.

cytology, and the older descriptive morphology are extremely difficult to harmonize with one another, even although, for various reasons, these difficulties are not apparent on a *prima facie* view. As soon as we do attempt such a synthesis we are confronted with contradictions which appear to rest on the fundamental biological antitheses. Instead of a unitary science we find something more approaching a "medley of *ad hoc* hypotheses."[13]

The belief that all the sciences were unifiable had been one of the cherished ideals of Enlightenment thought.[14] Heir to this thought, Auguste Comte in the mid-nineteenth century had articulated a positive philosophy that stated that the sciences went through three stages in their historical development: the theological, the metaphysical, and the positive. Examining the history of the sciences within Western thought, Comte postulated that each science was dependent for its grounding on previously existing sciences, so that biology was dependent on chemistry, which was dependent on physics, which was dependent on astronomy.[15] Though each science matured as it followed its own logic—revealed through close historical study of the science—there would still be an underlying unity to all the sciences. The progressive "growth" of knowledge within the framework often drew on the botanical metaphor of the branching "tree of knowledge."[16]

In the 1920s a belief in the unity of science was one of the fundamental tenets of the logical positivists of the Vienna Circle, the new heirs to Enlightenment thought. In the late 1920s and the 1930s Otto Neurath and Rudolf Carnap, members of this influential circle, and the Chicago-based Charles Morris spearheaded a movement to unify all the sciences that had emerged in the nineteenth and early twentieth centuries within a liberal worldview. Their Unity of Science Movement was based on their foundational belief that all the sciences—physical, biological, and social—were not only dependent *on*, but reducible *to*, physicalist terms that

---

[13] Ibid., p. 12.

[14] Belief in the unity of knowledge deeply structures Western thought. Plato discusses the unity of knowledge in his *Timaeus*; see *Timaeus and Critias*, trans. Desmond Lee (Harmondsworth, Middlesex: Penguin, 1965; repr. with rev. 1977).

[15] Comte was the thinker most responsible for promulgating the notion of a positivistic ordering of knowledge. For Comte, sociology—the science of society—emerging from physiology (biology) was to be the final science. See his *Cours de philosophie positive*, published in six volumes (Paris: Bachelier, 1830–42). Caron in his "'Biology' in the Life Sciences" introduces a discussion of Comte's use of the term "biology" and its close relation to physiology.

[16] Representations of the tree of knowledge proliferated during the Enlightenment. One of the most famous is included in Diderot and d'Alembert's *Encyclopédie*.

held one common method and protocol language. The movement not only upheld unification but, drawing from their intellectual progenitor Ernst Mach, stated that the unification of science had to take place by the elimination of metaphysics.[17] All true, legitimate sciences had to be purged of their metaphysical elements as they became grounded in fundamental axiomatic principles. The Unity of Science Movement swept intellectual circles in the late 1920s and the 1930s as international congresses were held, journals like *Erkenntnis* were established, and collaborative efforts to integrate knowledge within an *Encyclopedia of Unified Science* were launched.[18]

Woodger had been steeped in contemporary philosophical movements;[19] he was an active member and an outspoken advocate of the Unity of Science Movement, he attended congresses, and he corresponded with other members.[20] Consideration of the philosophical positions of progenitors of the logical positivists like Ernst Mach, of logical empiricist Bertrand Russell, and of mathematician-philosopher A. N. Whitehead and C. D. Broad was clear in Woodger's *Biological Principles*. But while he was to support strongly the Unity of Science Movement, and to urge the axiomatization and unification of biology, he was also to articulate explicitly a position for biologists that was antireductionist and anti-

[17] In an essay entitled "Ernst Mach and the Unity of Science," Philipp Frank summarized Mach's position on the unity of science in the following phrase: "He [Mach] proclaimed . . . *the unification of science by means of the elimination of metaphysics.*" Frank continued: "It is just this sentence that is the clue to the understanding of Mach's doctrine, of his papers, which seem to deal with so many subjects and such different fields of science. And it is just this program of Mach that we may adopt as the program of our 'Unity of Science Movement,' of our Congresses and of our Encyclopedia." Frank's essay was published in the official journal for the unity of science, *Erkenntnis*, and was later translated and republished in a collection of his essays. This quotation is on p. 243 of Robert S. Cohen and Raymond J. Seeger, eds., *Ernst Mach: Physicist and Philosopher*, Boston Studies in the Philosophy of Science, 6 (Dordrecht: D. Reidel, 1970), pp. 243–44.

[18] See Holton, "Ernst Mach and the Fortunes of Positivism in America."

[19] For a discussion on Woodger and his philosophical contacts, especially Rudolf Carnap and Max Black, see Pnina Abir-Am, "The Biotheoretical Gathering in England, 1932–38 and the Origins of Molecular Biology" (Ph.D. diss., Université de Montreal, 1983). See also Pnina G. Abir-Am, "The Biotheoretical Gathering, Trans-Disciplinary Authority and the Incipient Legitimation of Molecular Biology in the 1930s: New Perspective on the Historical Sociology of Science," *History of Science* 25 (1987): 1–70. For a discussion of the philosophical climate in Britain in the 1930s, see also A. J. Ayer, *Part of My Life* (London: William Collins and Sons, 1977).

[20] Woodger's role in the development of logical empiricism was recognized by Joergensen: see Joergen Joergensen, "The Development of Logical Empiricism," in Otto Neurath, Rudolf Carnap, and Charles Morris, eds., *Foundations of the Unity of Science* (Chicago: University of Chicago Press, 1970), II. After writing his *Biological Principles*, Woodger completed another book in 1937 on a related theme, *Axiomatic Method in Biology*; he also contributed a monograph for the *Encyclopedia of Unified Sciences* with the title, *The Technique of Theory Construction*. Woodger served on the advisory committee for the *Foundations of the Unity of Science: Toward an International Encyclopedia of Unified Science*.

physicalist.[21] The positivist drive to unify the sciences would, in fact, raise a series of complex and difficult issues for Woodger and other biologists.

While biology clearly had to grow out of its "metaphysical" stage of development and become a unified mature science,[22] full disciplinary unification—in a Machian and Unity of Science Movement sense—was actually dangerous to biology. If with unification there came a disciplinary reduction, then biology as a discipline was threatened with engulfment by the physical sciences.[23] Nor could life phenomena, which biology as the science of life sought to explain, be easily seen to be reducible to, or to obey, the laws of physics and chemistry, at least not without a great deal of discussion. For Woodger, who discussed and evaluated at length both vitalistic and mechanistic thinking in the history of biology, a middle ground—based solely on biological and not just physical principles—would have to be found in order to explain life phenomena. Both vitalism, which was too metaphysical, and mechanism, which drew too heavily on Newtonian physics, were inadequate for a mature science of life. For Woodger, the understanding of complex life phenomena would have to come solely from "observation and experiment" as articulated through exemplars in the physical sciences, rather than from any metaphysical and speculative considerations; but the biological principles he sought had to come from biology's own guiding logic rather than the logic of any physical science. If biology were to preserve its independent existence as a science, it could in some measure be dependent *on*, but could not be reduced fully *to*, physics and chemistry. If anything (and this was an increasing sentiment among biologists), physics and chemistry would have to accommodate biology.[24]

[21] Woodger's positivistic ordering was the following: physics, chemistry, biology, and psychology. Psychology and sociology were later combined into the larger category of the social sciences.

[22] Within the framework articulated by Comte the process of maturation and progression through these stages was inevitable.

[23] For this reason, many biologists in the 1920s and 1930s were to sympathize with the philosophical position articulated by A. N. Whitehead. In developing an antiphysicalist philosophical position, Whitehead had been drawing on his knowledge of biology to construct an organismic philosophy. See Ann L. Plamondon, *Whitehead's Organic Philosophy of Science* (Albany: State University of New York Press, 1979). According to Abir-Am, the Biotheoretical Gathering was inspired greatly by Whitehead and D'Arcy Thompson, See Abir-Am, "The Biotheoretical Gathering."

[24] A similar threat of disciplinary subsumption for the social sciences was of great concern to Otto Neurath as well. For a discussion of Neurath's views on this problem, see Danilo Zolo, *Reflexive Epistemology: The Philosophical Legacy of Otto Neurath*, trans. David McKie (Dordrecht: Kluwer, 1989). See especially chap. 5, "The Unity of Science as a Historico-sociological Goal: From the Primacy of Physics to the Epistemological Priority of Sociology."

According to some contemporary philosophers of biology, Woodger's influence on the development of the philosophy of biology was insignificant.[25] Yet given that no less a philosophical authority than the *Encyclopedia of Philosophy* is replete with Woodger's problematic under the heading "Biology," this is a difficult position to support.[26] However one wishes to situate Woodger within the history of the philosophy of biology, his position in the biological sciences accurately mirrored biologists' self-perceptions in the 1920s. Pointing out the immature state of biology, the lack of secure foundations with clearly articulated biological principles, and the disunified state of the biological sciences, Woodger's book became an urgent plea to axiomatize and unify the biological sciences and to bring them in line within the positivist ordering of knowledge—between physics and chemistry, on the one hand, and psychology and the other social sciences, on the other.

The call to axiomatize biology was echoed by J. S. Haldane, who faced

[25] See, for instance, David Hull's assessment of Woodger in one of the first review essays of the new field of the philosophy of biology that appeared in the late 1960s. According to Hull, Woodger "produced a body of work in the philosophy of biology too important not to include" in his survey, but which actually had little impact on biologists because they were not sufficiently versed in set theory and symbolic logic to understand his later work. See David Hull, "What Philosophy of Biology Is Not," *Journal of the History of Biology* 2 (1969): 241–68; quotation on p. 241. Because contemporary philosophers of biology (whose own history comes out of 1960s antireductionist movements in biology and antipositivist movements in philosophy) could not understand his later work, does not in any manner prove that Woodger was not an influential source for biologists and philosophers in the 1930s. From the numerous citations given to Woodger in a range of their books, biologists had frequently consulted Woodger's *Biological Principles* for understanding biology. Though few philosophers of biology would consider *The Encyclopedia of Philosophy* the definitive authorizing source for their field, it is widely consulted by general philosophers and other wider audiences; hence it serves (and has served) as an influential and authorizing source for broad audiences. Ernst Mayr had similar difficulty reading Woodger (personal communication).

[26] Paul Edwards, ed., *The Encyclopedia of Philosophy* (New York: Collier Macmillan, 1967), pp. 310–18. The entry for biology was written by Morton O. Beckner; the assessment of Woodger's *Biological Principles* at the conclusion of the entry reads as follows: "An influential and classical source of subsequent work in the philosophy of biology, partially Whiteheadian." Beckner also cited J. S. Haldane's *Philosophical Basis of Biology*, along with sources from authors like E. S. Russell, Ludwig von Bertalanffy, and Ernest Nagel. David Hull gave credit (but not without some criticism) to Morton Beckner's *The Biological Way of Thought*, which he stated "remains the single major contribution of a philosopher of biology in over a decade" ("What the Philosophy of Biology Is Not," p. 267). Clearly, philosophers of biology seem confused about their historical origins and to whom they give credit for what. For credit to Hull, see Michael Ruse, ed., *What the Philosophy of Biology Is: Essays Dedicated to David Hull* (Dordrecht: Kluwer, 1989). According to Ruse, "The success of this subject [the philosophy of biology] is due above all to the work and influence of one man: David Hull" (preface). According to Richard C. Lewontin, the most important influence on the philosophy of biology was exerted by Marjorie Grene. According to *Unifying Biology*, Ernst Mayr also played a key role. According to other philosophers, Richard C. Lewontin has played the greatest role by permitting numerous workers to train in his lab (chap. 2 discusses this).

some of the same issues head on in his 1931 book, *The Philosophical Basis of Biology*.[27] Haldane's discussion and endorsement of the need to ground biology in fundamental principles closely resembled Woodger's, an allegiance that Haldane explicitly favored. Both urged a critical examination of the fundamental logic of the biological sciences, which would be based on biology's own guiding principles, with biological explanation drawn on "facts" gleaned solely from observation and experiment. Only in this manner could biology become a legitimate and mature science, while preserving its independent status. Haldane explicitly stated that "biology must be regarded as an independent science with its own guiding logical ideas, which are not those of physics."[28] Both Haldane and Woodger denied the ability of exclusively mechanistic principles derived from Newtonian physics to explain life phenomena, and both highlighted the inadequacy of vitalistic thinking. Walking this fine line between mechanistic and vitalistic thinking in a manner that preserved the independent—and autonomous—status of biology, at the same time that it made possible scientific inquiry within a positivist philosophical framework, would prove to be *the* central problem of the twentieth-century biological sciences.[29] One thing was sure: by the 1930s, at least, biology was reaching a secure enough stage in its development for there to grow the conviction (in enough biologists' minds, at least) that the physical sciences would have to accommodate biology, and not the other way around. Haldane wrote: "That a meeting-point between biology and

[27] J. S. Haldane, *The Philosophical Basis of Biology* (London: Hodder and Stoughton, 1931). See also his earlier published version of his Gifford Lectures of 1927–28, *The Sciences and Philosophy* (London: Hodder and Stoughton, 1929), and *Philosophy of a Biologist*, 2d ed. (Oxford: Clarendon, 1936). In a supplemental section, written after the lectures that gave rise to his book in 1931, Haldane reviewed three books that had appeared at the same time, all of which discussed the fundamental principles of biology: Woodger's *Biological Principles*, E. S. Russell's *Interpretation of Development and Heredity*, and L. Hogben's *Nature of Living Matter*. While he disagreed mildly with Russell for upholding what he viewed as a standard "organismal" view of life, Haldane launched a full-blown attack against Hogben, who represented the strictly mechanistic conception of life that Haldane wished to avoid: "the foundations of this interpretation were entirely rotten. Moreover, physics was apparently almost entirely mechanistic, whereas fundamental mechanistic interpretation is now acknowledged to be impossible in physics. Professor Hogben stands bravely on a burning deck whence others have fled or are preparing to flee. We cannot but admire his courage" (pp. 164–65). For Haldane, the tension between vitalism and mechanism was remedied through a "holistic" view of life that emerged from the interaction of organism with environment.

[28] Haldane, *The Philosophical Basis of Biology*, p. 150.

[29] To contemporary philosophers of biology, the issue of the autonomy of biology is still considered central to any discussion of the philosophy of biology. Ernst Mayr, responding to Ernest Nagel and Carl Hempel, has contributed greatly to a viewpoint that gives autonomy to biology at the same time that biology becomes a legitimate science. See the discussion on "Biological Autonomy in the Post-*Sputnik* Biological Sciences" below.

physical science may at some time be found, there is no reason for doubt-
ing. But we may confidently predict that when a meeting-point is found,
and one of the two sciences is swallowed up, that one will not be biology."[30]

Nor was discussion of the relations between physics and biology con-
fined to the dialogue between Woodger and Haldane. The topic had
become especially popular for discussion by 1931. Joining Woodger and
Haldane, Joseph Needham, E. S. Russell, and J. D. Bernal, among other
biologists, faced the relation between physics and biology head on at a
special symposium on "The Historical and Contemporary Relationship
between Physics and Biology" at the Second International Congress of
the History of Science.[31] Later that year, at the centennial meeting of the
British Association for the Advancement of Science, General J. C. Smuts
delivered an honorary presidential address on the same subject. For
Smuts, who followed both developments in biology and developments in
physics, the danger of reduction could be avoided through a skillful combination of "emergent evolution" with the most recent developments in
quantum physics. Combining quantum physics and biology within what
he considered an organic worldview, he carried over quantum indeter-
minacy to the study of life. For Smuts (at least) the classic dilemma of
vitalism versus mechanism could thus effectively be dismissed since both
physics and biology were converging toward a new organic yet indeterminate scientific worldview.[32]

[30] This quotation, found on p. 33 of Haldane's *The Philosophical Basis of Biology*, is taken
from an address of 1908 reprinted in 1919 in *The New Physiology*. Haldane was also concerned to preserve the distinction between biology and psychology:

> In discussing the fundamental axiom of biology I have endeavoured to distinguish biol-
> ogy from the physical sciences and to illustrate the distinction. But the existence of con-
> scious behavior makes it necessary also to distinguish biology from psychology, the sci-
> ence, or rather great group of sciences or departments of knowledge, dealing with our
> experience when it is regarded as actually perceived and an expression of voluntary ac-
> tion. (Haldane, *Philosophical Basis*, p. 95)

The relationship between biology and psychology was not as great a concern to biologists,
since the threat of disciplinary engulfment through reduction was not as great a problem.
Psychologists were, however, concerned with a disciplinary reduction to biology. For this
discussion, see Nadine Weidman, "Of Rats and Men: Karl Lashley and American Psychology" (Ph.D. diss., Cornell University, 1994).

[31] See Pnina Abir-Am, "Recasting the Disciplinary Order in Science: A Deconstruction
of Rhetoric on 'Biology and Physics' at Two International Congresses in 1931," *Humanity
and Society* 9 (1985): 388–427. The conference met 29 June–3 July 1931. According to
Abir-Am the attendance (5,702 members) was more than double that of any other meeting
held between 1931 and 1935. For a record of the proceedings, see *Annual Reports* (London: The British Association for the Advancement of Science, 1931).

[32] J. C. Smuts, "The Scientific World-Picture of Today," *Annual Reports*, pp. 1–18; see
also *Holism and Evolution* (1926). Emergent evolution had become a popular doctrine in
the 1920s. It included followers like C. Lloyd Morgan and Samuel Alexander. See the

These two meetings catalyzed even further discussion. Such discussions, as historical sociologist Pnina Abir-Am has shown, fed an avante-garde British group of theoretically minded scientist-intellectuals. At the same time that they reflected on physics and biology, this self-designated "biotheoretical" group also thought about the history of science, philosophy, religion, and politics, following clearly the Enlightenment imperative to formulate a coherent worldview. Included in the "core group" were Woodger, Needham, and Bernal, along with embryologist Conrad H. Waddington and mathematician Dorothy M. Wrinch. The extended group included other notables like philosopher Max Black, physicist L. L. Whyte, and philosopher Karl Popper, as well as at least on one occasion polymath J. B. S. Haldane.[33]

Nor were they alone in exploring the relationship between physics and biology made apparent by the drive to unify the sciences.[34] Julian Huxley echoed both Woodger and Haldane in considering the maturation of biology and its relationship to physics and chemistry;[35] W. B. Turrill—a close colleague of Huxley—echoed Woodger's plea for an established logic or a "biologic" of the biological sciences;[36] and even in America,

---

discussion below on H. S. Jennings; for the connection between right-wing ideology and holistic, vitalistic, world-spirit philosophy, see nn. 34 and 44.

[33] See Abir-Am, "The Biotheoretical Gathering," for the whole list and the record of their attendance at meetings. It is especially hard to characterize J. B. S. Haldane. Originally he had trained in biochemistry with Gowland Hopkins at Cambridge. He then turned to genetics (both theoretical and applied) and mathematics in the 1930s. See the discussion below on Haldane and evolution. Later in life he became interested in exobiology. For a biography, see Ronald W. Clark, *The Life and Work of J. B. S. Haldane* (New York: Coward-McCann, 1969).

[34] Hans Driesch was another such biologist. In 1934 Driesch faced members of the Vienna Circle at the International Congress in Prague. Adhering to the strongest possible vitalistic philosophy, Driesch came under heavy fire from the Vienna Circle because he appeared to support nationalist philosophy and its quest for the "World Spirit" with the worst possible (in the Vienna Circle's view) brand of science. Peter Galison gives an account of how the Vienna Circle reacted to Driesch's plenary address in his Aufbau/Bauhaus article. The antithesis of the Drieschian vitalistic position is best exemplified by Jacques Loeb. For the "architects" of the evolutionary synthesis, Loeb—the most extreme of mechanists—would be too mechanistic and reductionistic. A middle road between the two positions would prove to be the most effective to give legitimacy and independence to the biological sciences. See the discussion below on Julian Huxley's political balancing act in the section entitled "*Evolution: The Modern Synthesis.*"

[35] Huxley wrote: "Every science arrives at a stage during which it makes its main broad contributions to practical human affairs. Biology is clearly on the verge of such a phase, while it is already over for physics and chemistry, and psychology and sociology cannot hope to reach it for perhaps another century" (Julian Huxley, "Biology and Physical Environment," in *What Dare I Think? The Challenge of Modern Science to Human Action and Belief* [New York: Harper and Row, 1931], p. 4).

[36] W. B. Turrill, "The Expansion of Taxonomy with Special Reference to Spermatophyta," *Biological Reviews* 13 (1938): 342–73.

William Morton Wheeler, reflecting on the difficulty of unifying a fragmented science, remarked that it might take "a few super-Einsteins" to
unify biology.[37]

Another biologist who would confront the same issues was Johns
Hopkins-based Herbert Spencer Jennings. Sympathetic to the Unity of
Science Movement, Jennings was acutely aware of the dangers of reduction facing biologists. As the threat of disciplinary reduction to physics
and chemistry loomed large, he constructed a powerful rationale for defending the independent existence of biology. Echoing J. C. Smuts, he
adopted a version of emergent evolution that could combine "the advantages of mechanism and of vitalism, dismissing the ineptitudes of each."[38]
At the same time his "doctrine of emergent evolution" would also lead to
his preferred scientific methodology of "radical experimentalism," which
also justified his emergent evolution. He wrote: "Radical experimentalism leads to emergent evolutionism, as emergent evolutionism leads to
radical experimentalism, they are indeed both." With truly emergent
properties that could be proved to exist through such radical experimentalism, moreover, biology could not be reduced to the physical sciences.
In a highly polemical, self-avowed "propaganda" article that was published in 1927 in *Science* from his retiring address of 28 December 1926
as chair of the zoological section of the American Association for the
Advancement of Science, Jennings laid bare his political drive to raise
biology to the status of an independent science. The connection between
his view of emergent evolution and biology was made transparent as he
triumphantly proclaimed:

> The doctrine of emergent evolution makes the biologist loyal to exper-
> imentation and observation in his own field of work, whatever is
> found in other fields. Courage and defiance sprout in his soul in place
> of timorous subservience to the inorganic. No longer can the biologist
> be bullied into suppressing observed results because they are not dis-
> covered nor expected from work on the non-living parts of nature. No
> longer will he feel a sense of criminality in speaking of relations that
> are obvious in the living, for the reason that they are not seen in the
> non-living. Biology becomes a science in its own right—not through
> rejection of the experimental method but through undeviating alle-

[37] W. M. Wheeler, *Essays in Philosophical Biology* (Cambridge, Mass.: Harvard University Press, 1939). Wheeler pointed out that the disunity in the biological sciences was indicative of the rich activity in those sciences.
[38] H. S. Jennings, "Diverse Doctrines of Evolution, Their Relation to the Practice of Science and of Life," *Science* 65 (1927): 19–25. Quotation on p. 25.

giance to it. The doctrine of emergent evolution is the Declaration of Independence for biological science.[39]

These authors were widely read by other biologists. A concern with the place of biology among the sciences; biology's maturity, status, and legitimacy as an independent science; the methodology of biology; the key role of observation and experiment; and a belief in the unity of science—all the critical considerations articulated by Woodger and echoed by Haldane, Jennings, and others were transmitted, in turn, to incipient biologists through the growing number of texts and textbooks of biology. Nor had these issues gained centrality solely from a one-way traffic of influence emanating from Woodger, Haldane, and the others. Woodger and Haldane (Woodger especially) had been responding to persistent concerns that accompanied the emergence of the biological sciences.[40] In writing their books, they were articulating and extending, but also codifying further, what was emerging as the disciplinary problematic[41] for the biological sciences.[42] These concerns, in turn, were increasingly becoming part of the established and received wisdom of the maturing biological sciences. By the 1930s biologists' belief in, and the drive toward, the unity of science and the unification of biology, as codified in the textbooks of biology, had been rendered tacit, unarticulated knowledge.[43]

As the driving force of positivism became more intense and as belief in the unity of science became part of the received wisdom of biology,

[39] Ibid., p. 21. William Morton Wheeler also adopted a version of emergent evolution, as did C. Lloyd Morgan and others.

[40] Woodger cites many of his biological contemporaries. Interestingly, Woodger had been reading and citing Arthur Dendy, *Outlines of Evolutionary Biology* (New York: Appleton, 1912). This book, which has been viewed as a minor text in the history of biology, may well prove to have been more influential on a wider audience.

[41] This can be defined as consisting of the inherited set of problems (which are discursively expressed), that are codified in the books of instruction or textbooks (which discipline students at the same time that they reproduce knowledge). These are accompanied by the tools, skills, and technologies that make possible the negotiated solutions of agreed-upon or common problems held by members of the culture. The disciplinary problematic bears some resemblance to Thomas Kuhn's disciplinary matrix.

[42] See H. W. Conn's "Biology: A New Science," in the introduction of *The Story of Life's Mechanism* (London: George Newnes, 1899). With regard to evolution he wrote:

A second conception, whose influence upon the development of biology was even greater, was the doctrine of evolution. It is true that the doctrine of evolution was no new doctrine with the middle of this century, for it had been conceived somewhat vaguely before. But until historical geology had been formulated, and until the idea of the unity of nature had dawned upon the minds of scientists, the doctrine of evolution had little significance. (p. 18)

[43] For one outstanding example, see Gairdner Moment, *General Biology for Colleges* (New York: Appleton-Century, 1942). The 1950 postsynthesis edition, in comparison to the 1942 presynthesis edition, reveals the more secure location of evolution as a legitimate area of scientific inquiry.

the threat of disciplinary subsumption became more imminent. Thus, with the peaking of positivism and the growing physicalism in the 1930s, there also arose biological movements that, in reacting to or adjusting to a positivist framework, made attempts to preserve the independence of biology. These movements, variously characterized as "holistic," "organicist," "emergentist," or "organismal," upheld in some manner the view that there were independent properties to life.[44] The independence of life made possible the independent existence of the science of life, biology. The rise of an avowed "neovitalistic" way of thought was explicitly articulated by J. Arthur Thomson and Patrick Geddes, who most strongly supported the unity of science in their textbook, *Life: Outlines of General Biology*.[45] Explicitly discussing and visualizing the orderly arrangement of the sciences (see fig. 1 and accompanying explanation), they stated that each of the sciences rested on a base of other sciences, the most universal of which was logic. Each successive science "retains its own distinctiveness" through a "fresh 'Emergence.'" Each field deserved full investigation for "its own sake, and for its services to others." In this way "The Unity of Science" was thus to be "realized."[46]

Though justifying the independence of biology and the independence of life phenomena, these biological movements could still be accused of being too speculative, mystical, and metaphysical, and therefore of con-

[44] The movements were simultaneous with a resurgence of conservative ideology. Henri Bergson—the most prominent philosopher in the early twentieth century—was heavily supported by conservative groups. Bergson and Bergsonianism were part of a larger "revolt from mechanism" that had accompanied the rise of positivistic philosophy and the revival of the occult in France. See R. C. Grogin, *The Bergsonian Controversy in France, 1900–1914* (Calgary: University of Calgary Press, 1988), for a discussion of the cultural milieu surrounding Bergson and Bergsonianism and for more full discussion of how Bergson was received by different groups. The logical positivists and the "architects" of the evolutionary synthesis may well be viewed as silencing Bergson. Fisher, Haldane, and especially Huxley, whose early book, *The Stream of Life*, could be seen to espouse a moderate Bergsonianism, had all been responding to Bergson in the 1920s; see Julian Huxley, *The Stream of Life* (New York: Harper and Row, 1927). Huxley had been heavily influenced by Bergsonian philosophy as early as 1912. See his first book, *The Individual in the Animal Kingdom* (Cambridge: Cambridge University Press, 1912), and the comments of praise for Bergson: "It will easily be seen how much I owe to M. Bergson, who, whether one agrees or not with his views, has given a stimulus (most valuable gift of all) to Biology and Philosophy alike" (p. vii). The other life-philosopher whom Huxley cited was Nietzsche.

[45] See J. Arthur Thomson and Patrick Geddes, *Life: Outlines of General Biology*, 2 vols. (London: Williams and Norgate, 1931). A 1925 edition of an earlier book by Geddes and Thomson entitled *Biology* had been cited by Woodger in his *Biological Principles*. Thomson and Geddes recommended Whitehead for further reading under the heading "Biology and Philosophy"; in the addendum to the second volume (p. 1499), they recommended Woodger's *Biological Principles* for further reading. Thomson had been a proponent of Bergson. He also supported Hans Driesch for the Gifford Lectureship.

[46] See the "Explanation of 'Sciences in General' End-Paper," in Thomson and Geddes, *Life*. See also fig. 1.

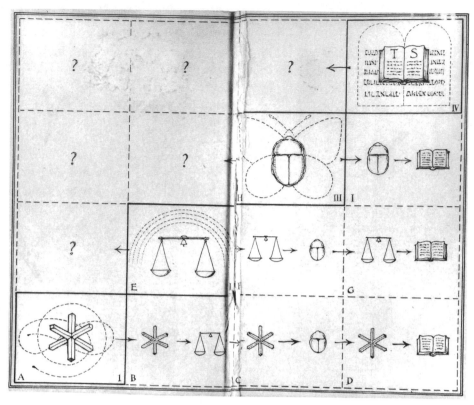

FIGURE 1: "Sciences in General" endpaper from *Life: Outlines of General Biology* and accompanying "Explanation of 'Sciences in General' End-Paper" (J. Arthur Thomson and Patrick Geddes, *Life: Outlines of General Biology*, vol. 1 [London: Williams and Norgate, 1931]).

(A) Lowest on the left, the field of pure Mathematics, is marked by axes in three dimensions.

The space or step above this towards the right, indicates the field of the Physical Sciences, symbolized by the Balance.

In the next space (above to right) the field of Biology, indicated by the Scarabaeus.

And in the last space (above to right) the field of Sociology—indicated by the Book; and marked "T" and "S" for its "Temporal and Spiritual" elements, more or less conspicuous in each and every form of Social Life, and its Heritage (and Burden).

(B)   Descending in reverse order, note, behind the Book, for Sociology, Moses' Tables of Commandments, as old symbol for Ethics. Beyond the Scarab of Biology spreads the butterfly (*Psyche*) of Psychology. Above the Balance for Physical Science, the Rainbow (*Iris*) of Esthetics: and around the Mathematical Axes the swirl of Logic, as most universal of all sciences. This descending series of subjective sciences is in Plato's order; of "Good, True, and Beautiful"; and is thus complemental to the previous ascending series—that from Aristotle to Bacon, Comte to Spencer, and modern scientific workers generally.

(C)—Again in ascending order, note that each main science, shown on its step, is preliminary to the succeeding one—and also extends on its own level below it. Thus Mathematics primarily subserves the Physical Sciences; yet also Biology, as Biometrics; and Social Science, as Statistics. The Physical Sciences similarly underlie Biology (as Bio-Mechanics, Bio-Physics, and Bio-Chemistry); and even the Social Sciences also, since all in physical environments. Biology underlies Social Science; since the Social life of Region and State, City and Citizen, are all biologically conditioned (as by Physiology and Hygiene, with Heredity, Eugenics, etc.).

   Yet each succeeding science retains its own distinctiveness-as fresh "Emergence". Hence with the clear foundation of Sociology, on "the preliminary sciences", their respective underlying contributions were defined as so many "*legitimate Materialisms*". Only when the needed preliminary contribution—to each main field or fields above— is mistakenly assumed sufficient to supersede it or them (as too often by their most active cultivators, and thus their readers) do "illegitimate Materialisms" arise. Bio-Physicists and Bio-Chemists for Biology, and Hygienists and Eugenists for Sociology, are alike often liable to such errors.

(D)—the interrogation-marks in the six spaces otherwise vacant (to the left of the diagonally ascending stair of the main sciences) indicate the fields for inquiry into the suggestive contributions of each higher science to its so far "preliminary sciences". Hence then for the evocations from Social Life and Science to Biology and Psychology. Biology has similarly aided Physical Science; and Psychology Esthetics; and similarly Physical problems evoke Mathematical methods, and even advance Logic.

   Such inquiries (see Appendix II) have been termed "legitimate transcendentalisms"—and are to be distinguished from "illegitimate" ones;—as of some philosophers, with inadequate knowledge of the sciences, and sometimes by their specialists as well. Thus each and every one of the sixteen fields of this Graphic needs and rewards full investigation, alike for its own sake, and for its services to others. The Unity of Science has thus to be realised.

stituting bad science within a positivist framework. The most effective manner to unify biology, accommodate positivistic science, and preserve the independence of biology (the same fine line walked by Woodger, Haldane, and the others) would lie in the elaboration of a position more reminiscent of Jennings. Evolution, purged of unacceptable metaphysical elements, would function as the phenomenon that could make biology an "autonomous" science, at the same time that it served as the "unifying principle" that Woodger and others had sought. Evolution would be made to "lift" biology above the physical sciences at the same time that it "bound" the fractured biological sciences. But first, evolution—threatened with extinction—had to be made into a legitimate science. One outcome of the "evolutionary synthesis" would be the making of such a science.

The Decline of Natural History and Evolutionary Studies

Within a positivist theory of knowledge,[47] the branches of biology that suffered most from critical methodological scrutiny were the natural history-oriented sciences. Especially vulnerable were evolutionary studies. Not only did the study of evolution reek of metaphysical and vitalistic elements—witness the popularity of such views as directed evolution, Lamarckism, emergent evolution, and the numerous other agents of evolution outlined by Darwin and his heirs—but it also defied the great method for grounding positivistic claims to knowledge: experimentation. Nor was there much direct empirical, observational evidence for evolution, which was a historical science.

The rise of genetics—the first unquestionably mechanistic, materialistic, law-like (e.g., with the existence of Mendel's Laws), and experimental biological science—at the turn of the century,[48] combined with the increasing experimentalization of older established biological sciences like embryology and physiology, led to a period of turbulence for natural history.[49] As academic institutions, especially in the United States, were

[47] I am using the term "positivist" here to note the currents discussed by Holton, "Ernst Mach and the Fortunes of Positivism in America."
[48] Unlike the "-ologies," which were language-oriented or descriptive sciences, genetics was an "-ics" word, meant to emulate and resemble physics and other exact sciences. For a discussion of the privileged location of genetics in the biological sciences, see V. B. Smocovitis, "Talking about Sociobiology," *Social Epistemology* 6 (1992): 219–30.
[49] See Garland E. Allen, *Life Science in the Twentieth Century* (New York: Wiley, 1975). See also Elizabeth Gasking, *The Rise of Experimental Biology* (New York: Random House, 1970). Jane Maienschein has also pointed out that evolution "failed to expand" in the same

engaged in reforming and restructuring their curricula and instituting new departments, experimental sciences were increasingly favored over natural history or descriptive sciences. By the late 1930s natural history, which included study of evolution, was threatened with extinction, especially on university campuses. In the United States, courses of instruction that included substantive discussion of evolutionary studies were increasingly unavailable at places like Harvard, where course catalogues indicate a proliferation of courses on experimental biological sciences.[50] General biological textbooks devoted less and less space to evolution, which often appeared as an afterthought in the final chapter.[51] Even histories of biology—a new historical genre—such as that of Nordenskiöld and Rádl indicated that evolution was thought to be a dead or dying subject.[52] Fewer and fewer scientific periodicals were willing to publish articles on evolutionary subjects, which were becoming unpopular with readers. A quick survey of articles in the leading forum for naturalists in America, the *American Naturalist*, indicates the extent to which nonevolutionary geneticists, physiologists, and embryologists had infiltrated what had long been the leading naturalists' journal. In the mid-1930s there was even an attempt to take over the *American Naturalist* and turn it into a journal for genetics.[53]

---

way (professionally and academically) that other fields had during this expansive phase in the history of American biology (letter to author from Jane Maienschein, 19 June 1995).

[50] Course catalogues at key American institutions like Harvard University during the 1897–1967 period are a powerful indicator of the history of evolution in America. The first course at Harvard University with a sole emphasis on evolution was given in 1908–9. Entitled Zoology 20d, *Investigation of the Factors Involved in Evolution*, this was not a regular course of instruction, but was listed under "Courses of Research." It was given by W. E. Castle at the Bussey Institution, and was offered for only one year. Nearly a twenty-year interval separated this course from the next course with a sole emphasis on evolution: Biology 112a, *Problems of Evolution*, which was offered in the academic year 1937–38. This course was intended primarily for graduate students and was taught by Edward Murray East, a geneticist at the Bussey Institution. East died in 1938, and the course was not offered further. Ernst Mayr taught Biology 147 (later changed to 247), *Systematics and Evolution*, in 1954–55. The first course entitled *Evolutionary Biology* was Biology 144 taught by E. O. Wilson primarily to undergraduates. Ernst Mayr taught Biology 248, *Principles of the Evolutionary Theory*, for graduate students in the same year.

[51] This includes general zoology textbooks and botany textbooks.

[52] Erik Nordenskiöld, *The History of Biology* (New York: Alfred A. Knopf, 1928; trans. from Swedish ed. of 1920–24). Interestingly, Nordenskiöld was more concerned with colloid chemistry than evolution. Emmanuel Rádl, *The History of Biological Theories*, trans. E. J. Hatfield (Oxford: Oxford University Press; London: Humphrey Milford, 1930; orig. pub. as *Geschichte der biologischen Theorien in der Neuzeit*, 2 vols. [Leipzig: Engelmann, 1907–8; rev. ed. 1913]).

[53] The *American Naturalist* had been privately owned by Jaques Cattell Press. In the mid-1930s the journal began to accommodate the market not only for experimental sciences like genetics and physiology but also for sociology and psychology, and thus devoted less and less space to traditional naturalist articles. The move to found a society for the study of evolution was part of an increasing need that naturalist-systematists felt for creating an information service, with a journal to publish work that would otherwise have gone

Morever, funding for evolutionary research was nearly impossible to obtain. The enlightened despot of biological finance, Warren Weaver at the Rockefeller Foundation, was unsympathetic to nonexperimental and nonmedical fields.[54] Those most unsympathetic to natural history and the descriptive study of evolution were experimental biologists and embryologists like E. G. Conklin, and some geneticists like T. H. Morgan, who were impatient with what they saw as an unrigorous science and overly metaphysical area of inquiry. "'Naturalist' was a word almost of contempt with him, the antonym of 'scientist,'" Morgan's own apprentice, Theodosius Dobzhansky, later recalled.[55] Even though Morgan was thought to have softened his harsh early stance on existing study of evolution by the 1930s, his 1932 book, *The Scientific Basis of Evolution*, implied that there had been little rigorous scientific basis for evolution before Morgan himself.[56]

---

to the *American Naturalist*. Documents deposited at the Library of the American Philosophical Society indicate an increasing dissatisfaction with existing journals, beginning in the 1930s and extending to the early 1940s. My perspective has been reinforced by Ernst Mayr (letters from Ernst Mayr to author, 27 February 1989 and 15 August 1989). For a documentary history of the Society for the Study of Evolution (SSE) and the founding of a journal expressly devoted to evolution, see Vassiliki Betty Smocovitis, "Organizing Evolution: Founding the Society for the Study of Evolution (1939–1950)," *Journal of the History of Biology* 27 (1994): 241–309. For a historical account of organizational activities in evolution centering on the Committee on Common Problems of Genetics, Paleontology, and Systematics, see Joseph Allen Cain, "Common Problems and Cooperative Solutions: Organizational Activity in Evolutionary Studies," *Isis* 84 (1993): 1–25. See also discussion below.

[54] Both Ernst Mayr and G. Ledyard Stebbins were unable to obtain funds from the Rockefeller Foundation (letter from Ernst Mayr to author, 15 August 1989; letter from G. Ledyard Stebbins to author, 4 May 1989). This was to become even more problematic after the "discovery" of DNA. See Mayr's impassioned plea to continue the support of "classical" biology in the face of the "new" biology in the discussion below. For a discussion of the Rockefeller Foundation's commitment to support research that would bring together the physical and biological sciences, see Pnina G. Abir-Am, "The Assessment of Interdisciplinary Research in the 1930s: The Rockefeller Foundation and Physico-chemical Morphology," *Minerva* 26 (1988): 153–76. See also Robert Kohler, *From Medical Chemistry to Biochemistry* (Cambridge: Cambridge University Press, 1982); and Lily E. Kay, *The Molecular Vision of Life* (New York: Oxford University Press, 1993).

[55] Dobzhansky also added: "Yet Morgan himself was an excellent naturalist, not only knowing animals and plants but aesthetically enjoying the observing of them." This and other contradictory elements in Morgan's weltanschauung are discussed in Theodosius Dobzhansky, "Morgan and His School in the 1930s," in Ernst Mayr and William B. Provine, eds., *The Evolutionary Synthesis: Perspectives on the Unification of Biology* (Cambridge, Mass.: Harvard University Press, 1980), pp. 445–52. Quotation on p. 446. Dobzhansky later wrote: "To serve the function that Morgan had assigned to it, biology had to be free of any taint or suspicion of philosophical idealism, vitalism, or teleology. It had to be strictly reductionistic. Biological phenomena had to be explained in terms of chemistry and physics. Morgan himself knew little chemistry, but the less he knew the more he was fascinated by the powers he believed chemistry to possess." Quotation on p. 447.

[56] T. H. Morgan, *The Scientific Basis of Evolution* (New York: W. W. Norton, 1932). Morgan's critique of existing evolutionary theory, the inadequacies of paleontology for

The decline in scientific standing of evolutionary studies was felt widely at the time. Where it was close to devastating was in Darwin's own homeland. Julian Huxley coined the phrase "the eclipse of Darwin" to highlight the downturn in his *Evolution: The Modern Synthesis.* Historians and biologists have long recognized this period in the history of biology, but few have interpreted the phenomenon in a disciplinary manner.[57] The eclipse of Darwin referred not just to the demise of Darwin's theory of natural selection—natural selection was unsatisfying and was never widely or exclusively accepted, even by the most fervent of Darwin's followers—but also to the demise of natural history and evolutionary studies, the fields that Darwin had represented to his heirs. So long as the agents of evolution were unquantifiable, and evolution remained a nonexperimental practice, evolution was bound to be a speculative, unrigorous, and ultimately endangered science. Writing about the sad state of Darwinism at the turn of the century, Vernon L. Kellogg explicitly stated that Darwinism was undergoing methodological scrutiny and was coming under fire from "German biologists" and experimental biology itself. In his introductory discussion of the "Death-Bed of Darwinism" he wrote: "there is going on a most careful re-examination or scrutiny of the theories connected with organic evolution, resulting in much destructive criticism of certain long-cherished and widely held beliefs, and at the same time there are being developed and almost feverishly driven forward certain fascinating and fundamentally important new lines, employing new methods, of biological investigation. Conspicuous among these new kinds of work are the statistical or quantitative study of variations and that most alluring work variously called developmental mechanics, experimental morphology, experimental physiology of development, or, most suitably of all because most comprehensively, experimental biology." Kellogg continued: "Now this combination of destructive critical activity and active constructive experimental investigation has plainly resulted, or is resulting, in the distinct weakening or modifying of certain

---

providing explanatory accounts, and the importance of his own (rather self-serving) experimental work in heredity and evolution were discussed in his Louis Clark Vanuxem Lectures delivered at Princeton University in 1916. See *A Critique of the Theory of Evolution* (Princeton: Princeton University Press, 1916). For a discussion of the development of Morgan's views on evolution, see G. E. Allen, "Thomas Hunt Morgan and the Problem of Natural Selection," *Journal of the History of Biology* 1 (1968): 113–39; and idem, *Thomas Hunt Morgan: The Man and His Science* (Princeton: Princeton University Press, 1978).

[57] See, for instance, Peter Bowler, *The Eclipse of Darwinism: Anti-Darwinian Evolutionary Theories in the Decades around 1900* (Baltimore: Johns Hopkins University Press, 1983); and Ernst Mayr, *The Growth of Biological Thought* (Cambridge, Mass.: Belknap Press of Harvard University Press, 1982).

familiar and long-entrenched theories concerning the causative factors and the mechanism of organic evolution. Most conspicuous among these theories now in the white light of scientific scrutiny are those established by Darwin, and known, collectively, to biologists, as Darwinism."[58]

The spread of positivist philosophy was especially rapid in the United States. As historian of science Gerald Holton has argued, philosophical currents from Mach were established in a range of disciplines and instituted officially at key American intellectual centers of research and instruction like Harvard University.[59] In the 1920s the influence of Jacques Loeb, the arch-mechanistic materialist and disciple of Mach, had been widely felt by physiologists, biologists, and especially some psychologists.[60] At Harvard noted physiologist W. J. Crozier had been an especially devoted disciple of Loeb. For Crozier, evolution, which lacked rigorous experimental methods, could not be considered a proper science. Embodying this positivistic and antievolution attitude, Crozier asserted to his numerous students in his introductory biology course at Harvard: "Evolution is a good topic for the Sunday supplements of newspapers, but isn't science: You can't experiment with two million years!"[61]

As positivistic currents structured modern thought,[62] nonexperimental sciences fell to accusations of being inexact, imprecise, and unrigorous. This tension between experimentalists and nonexperimentalists just around and after the turn of the century has long been recognized by historians of biology Garland Allen and Ernst Mayr.[63] Genetics, physiol-

---

[58] Vernon L. Kellogg, *Darwinism To-day* (New York: Henry Holt, 1907), pp. 1–2.

[59] Holton, "The Fortunes of Mach in America."

[60] See Jacques Loeb, *The Mechanistic Conception of Life* (Cambridge, Mass.: Harvard University Press, 1964 repr. ed.). For a biography of Loeb, see Philip J. Pauly, *Jacques Loeb and the Engineering Ideal in Biology* (New York: Oxford University Press, 1987). For a full examination of the relationship between behaviorism and logical positivism, see Laurence D. Smith, *Behaviorism and Logical Positivism: A Reassessment of the Alliance* (Stanford: Stanford University Press, 1986).

[61] G. Ledyard Stebbins, unpublished autobiographical manuscript, "Getting There Is Half the Fun," p. 10. Stebbins had been a student of Crozier. It was Crozier's negative comments directed against evolution that fueled Stebbins' interest in evolution. B. F. Skinner and Gregory Pincus were also in the same class.

[62] This included the progenitors of the logical positivists and a loose assortment of philosophies that fed the positivistic attitude. For one historical exploration, see chap. 2, in Smith, *Behaviorism and Logical Positivism*.

[63] See Garland Allen, "Naturalists and Experimentalists: The Genotype and the Phenotype," *Studies in History of Biology* 3 (1979): 179–209. The conflict between naturalists and experimentalists is the basis for the "Allen" thesis. The applicability of this thesis has been a contentious issue for historians of biology. My own position strongly upholds the Allen thesis. Within a positivist framework, nonexperimental sciences would be greatly suspect. Within this framework, the conflict between geneticists and naturalists, which Ernst Mayr has recognized, is also understandable. Genetics, being experimental (based on breeding practices) and an increasingly established mechanistic and materialistic science, would

ogy, embryology, and other experimental sciences leaned to one side, while natural history and evolutionary studies (historical, nonexperimental, and unquantifiable sciences with little direct empirical evidence) leaned to the other side.[64] If observation and experiment—essential positivistic criteria for legitimate science—were not part of the methodologilcal apparatus of the discipline, the discipline came under fire, and the tensions only increased with time.[65] By the late 1920s—as Woodger and the others were assessing the status of the biological sciences—the lack of rigor and grounding in evolutionary studies had become even more acute. The 1930s soon witnessed the emergence of a group of biologists who were to serve as bridge builders and "architects" by adopting and transferring methodologies from the physical sciences to genetics, systematics, botany, and paleontology (and other related sciences) to make the study of evolution a more rigorous, legitimate science. In so doing they constructed a unified and autonomous *science* of biology. "Modernizing" evolution, they were also to preserve the naturalistic, Darwinian tradition that had gone into decline.

THEORY AND EXPERIMENT IN EVOLUTION

The critical moment for evolutionary studies—and the unification of biology—came with the successful adoption of experimentation in evolutionary practice through the work of mathematical modelers working in conjunction with field naturalist-systematists. This took place during the

---

be favored over nonexperimental and descriptive sciences. Other laboratory-oriented experimental sciences like cytology and embryology would also preserve their "rigorous" science status over descriptive, natural history-oriented evolutionary studies. See also Smith, *Behaviorism and Logical Positivism.*

[64] Vernon Kellogg's 1907 *Darwinism To-day* supports this argument. Kellogg concluded with the following thoughts:

We are ignorant; terribly, immensely ignorant. And our work is, to learn. To observe, to experiment, to tabulate, to induce, to deduce. Biology was never a clearer or more inviting field for fascinating, joyful, hopeful work. To question life by new methods, from new angles, on closer terms, under more precise conditions of control; this is the requirement and the opportunity of the biologist of to-day. May his generation hear some whisper from the Sphinx!" (p. 387)

[65] These criticisms were leveled not only at the biological sciences, but at other disciplines as well. The discipline of history, for instance, emerged as a legitimate area of inquiry only when scientific—and positivistic—standards began to be adopted in the late nineteenth century. "Observation," and eventually—through quantitative modeling in the 1960s—"experiment," were adopted by historians. History in this manner became a "social science." See, for instance, the quantitative historical work that boomed in the 1960s. For a recent annotated bibliographic survey of most of this literature, see Stephen R. Grossbart, "Quantitative and Social Science Methods for Historians," *Historical Methods* 25 (1992): 100–120.

celebrated interactions between mathematical theorists and field biologists—in England, between R. A. Fisher and E. B. Ford, and in America, between Sewall Wright and Theodosius Dobzhansky. To be sure, there had been other attempts at experimentation in evolution,[66] including J. W. Heslop Harrison's work on melanism in *Lepidoptera*, H. C. Bumpus's natural experiments with sparrows, W. F. R. Weldon's work with crabs, and even the much-celebrated experiments of Weismann and Darwin himself,[67] but none of these attempts had led successfully to the most "objective" type of knowledge:[68] the attachment of numbers to nature within a mechanistic and materialistic framework.[69] Only after the

[66] In the early twentieth century the move to experimentalize evolution was institutionalized officially by the founding of an experiment station expressly devoted to this end: in 1904 the Carnegie Institute of Washington supported the founding of the Station for Experimental Evolution at Cold Spring Harbor, Long Island, New York. The first director was C. B. Davenport.

[67] J. W. Heslop Harrison, "The Induction of Melanism in the *Lepidoptera* and Its Evolutionary Significance," *Nature* 119 (1927): 127–29; H. C. Bumpus, "The Elimination of the Unfit as Illustrated by the Introduced Sparrow, *Passer domesticus*," in *Biological Lectures from the Marine Laboratory, 1898* (Boston: Ginn, 1899), pp. 209–26; W. F. R. Weldon, "An Attempt to Measure the Death Rate Due to the Selective Destruction of *Carcinus Moenas* with Respect to a Particular Dimension," *Proceedings of the Royal Society of London* 57 (1895): 360–79.

[68] For a recent account of the historical relationship between quantification in science and the construction of scientific objectivity, see Theodore M. Porter, *The Pursuit of Objectivity in Science and Public Life* (Princeton: Princeton University Press, 1995).

[69] Additional support comes from examination of a book meant to introduce readers to "great experiments in biology," in which the ordering of the topics roughly reflects chronological experimentalization (and also legitimation): The Cell Theory (Cytology), General Physiology, Microbiology, Plant Physiology, Embryology, Genetics, and lastly Evolution. Four examples of classic experiments in evolution are given: selections from C. Darwin and A. R. Wallace (not experimental), G. H. Hardy (mathematical, but not experimental), N. H. Horowitz (experimental but biochemical), and Th. Dobzhansky (experiments with natural populations of organisms). The ordering as well as the selections justify the sense that it is Dobzhansky's work with natural populations that forms the critical moment of the experimentalization of evolution. See Mordecai L. Gabriel and Seymour Fogel, eds., *Great Experiments in Biology* (Englewood Cliffs, N.J.: Prentice Hall, 1955). They write in the introduction to the abridged version of a paper by Dobzhansky entitled "Adaptive Changes Induced by Natural Selection in Wild Populations of Drosophila," *Evolution* 1 (1947): 1–16:

> It commonly used to be said, as a defense against critics of the evolution theory, that natural selection was such a slow process that it was not to be expected that evolutionary changes would be perceptible within the span of a human lifetime. Such pessimism turned out to be unjustified, however. By the choice of organisms with a sufficiently short generation time, such as *Drosophila* and bacteria, it has proven possible not only to detect evolutionary change in the laboratory but to measure its rate and study the effective forces in quantitative terms. An admirable example of this type of research is the following study by Dobzhansky, an outstanding contributor to the modern synthesis of genetics and evolution. So decisively have recent genetic studies vindicated Darwinism to all but a few dissidents that there is irony in the recollection that the early geneticists believed that the facts of Mendelian inheritance cast serious doubts upon the Darwinian hypothesis. (p. 301)

variables of natural selection, genetic drift, and mutation were articulated by mathematicians in the early twentieth century, to formulate the "Hardy–Weinberg Equilibrium Principle,"[70] and only after there was agreement that these were the legitimate variables in evolution,[71] could they be measured and made to work in natural populations. By demonstrating how these variables could work in natural populations, the newly constructed mathematical models—systems of interacting variables—of R. A. Fisher, J. B. S. Haldane, and Sewall Wright had utility for field-oriented naturalists needing a workable methodology. They also helped begin to purge belief in a variety of "alternative mechanisms" for evolution, by making natural selection viable once again.[72]

These mathematical models had not arisen de novo, however. Fisher, Haldane, and Wright had been issued detailed prescriptives by biologists in the field as well as biologists on the farm, and all three had views of population structures in mind as they began to articulate their models and transmit their representations to field biologists Ford and Dobzhansky.[73] Historians know little about the nature of the dialogue be-

---

[70] The Hardy–Weinberg Equilibrium Principle describes the conditions under which evolutionary equilibrium is maintained—that is, the conditions under which changes in gene frequencies or genotypes *do not* take place. For a history of mathematical population genetics, see William B. Provine, *The Origins of Theoretical Population Genetics* (Chicago: University of Chicago Press, 1971); idem, "The Role of Mathematical Population Geneticists in the Evolutionary Synthesis of the 1930s and 1940s," *Studies in History of Biology* 2 (1978): 167–92. The Hardy–Weinberg Equilibrium Principle now goes by the name of Castle–Hardy–Weinberg, acknowledging the work of W. E. Castle. "Gene" frequencies have now been replaced by the more precise "allelic" frequencies.

[71] See William B. Provine, *Sewall Wright and Evolutionary Biology* (Chicago: University of Chicago Press, 1986), for a discussion of the negotiations leading to agreement over the relative importance of these variables.

[72] See Provine, *Origins of Theoretical Population Genetics*; idem, "Role of Mathematical Population Geneticists"; Sewall Wright, "Evolution in Mendelian Populations," *Genetics* 16 (1931): 97–159; idem, "The Roles of Mutation, Inbreeding, Crossbreeding, and Selection in Evolution," *Proceedings of the Sixth International Congress of Genetics* 6 (1932): 356–66; J. B. S. Haldane, "A Mathematical Theory of Natural and Artificial Selection," *Proceedings of the Cambridge Philosophical Society* (1924–32); idem, *The Causes of Evolution* (1932; repr. Ithaca: Cornell University Press, 1966); R. A. Fisher, *The Genetical Theory of Natural Selection* (Oxford: Oxford University Press, 1930).

[73] See, for instance, the work of naturalist-systematists who had been working independently of the mathematicians to study geographic variation. Many of these workers had upheld some form of Lamarckian inheritance or mutationism. These include Francis Sumner on the geographic races of the deer mouse: F. B. Sumner, "Genetic, Distributional and Evolutionary Studies of the Subspecies of Deer-mice (*Peromyscus*)," *Bibliographia Genetica* 9 (1932): 1–106. See also the work of Richard Goldschmidt on the geographic races of *Lymantria dispar*: Richard Goldschmidt, "*Lymantria*," *Bibliographia Genetica* 11 (1934): 1–186. Botanists had been especially active in studies of variation in nature: these individuals include Göte Turesson, Erwin Baur, and Edgar Anderson. See Vassiliki Betty Smocovitis, "Botany and the Evolutionary Synthesis: The Life and Work of G. Ledyard Stebbins" (Ph.D. diss., Cornell University, 1988). See the discussion below on those naturalists who upheld nonadaptive evolution.

tween Fisher and Ford, who engaged each other in person and left little correspondence behind, but historian William Provine has mapped out the manner and proximity of the interaction between their American counterparts, Dobzhansky and Wright.[74] The interaction between Wright and Dobzhansky did not lead to a "miraculous correspondence" of mathematics and nature, but was the product of an ongoing dialogue between workers at the desk and workers in the field and on the farm. Both "theorist" and "experimentalist" were engaged in practical attempts to solve agreed-upon problems within evolutionary studies, the most central of which was the determination of the agents responsible for evolution in natural populations of organisms. The mathematical models of Wright and the organismal models of Dobzhansky became developed through the interactive efforts of both into coadapted heuristics, which successfully led to the attachment of numbers to nature.[75]

The quantification of evolution[76]—the attachment of numbers to "nature"—and the growing measurability and testability of natural selection within a populational framework were part of a process that would eventually lead to general support for natural selection as the primary mechanism of evolution.

THE "PURIFICATION" OF EVOLUTION AND THE "UNIFICATION" OF BIOLOGY

While organic evolution had been perceived as a causo-mechanical explanation for organic change,[77] no real, indisputable causo-mechanical explanation had been satisfactorily demonstrated. For readers of *Origin of Species*, natural selection was viewed as one of several "agents" (i.e., a causo-mechanical "agent") of evolution. Belief in the agency of natural selection was the outcome of the manner in which Darwin had structured his argument for natural selection, arguing from analogy with arti-

[74] See E. B. Ford, *Mendelism and Evolution* (London: Methuen, 1931); Provine, *Sewall Wright*, pp. 327–65.

[75] Dobzhansky actively sought a model organism, *Drosophila pseudoobscura*, to fit Wright's schemes. *Drosophila melanogaster*, the familiar tool of the Morgan school's genetics, did not display phenomena that could be made to work with Wright's evolutionary models. Provine recounts the story of how Dobzhansky became a "victim" of "scientific schizophrenia" until Robert D. Boche gave him *Drosophila pseudoobscura* (Provine, *Sewall Wright*, p. 332). For the importance of *Drosophila* as instrumental organism in the history of classical genetics, see Robert Kohler, *Lords of the Fly*: Drosophila *Genetics and the Experimental Life* (Chicago: University of Chicago Press, 1994).

[76] Mutation, migration, population structure, and systems of mating, as well as random genetic drift, became quantifiable and measurable.

[77] Causo-mechanical is the precise phrase used by Vernon Kellogg in his 1907 book.

ficial selection. Although natural selection did not take place through the "hand" of a selector, the view of natural selection as agent still had built into it a degree of purposiveness, of teleology, and therefore it had a "metaphysical" taint. This was at least one reason for the growing decline in evolutionary studies.

With the work of the modelers and the adoption of experimental and quantitative methods, the view of natural selection as agent began to diminish, so that by the 1930s natural selection took on a causo-mechanical existence. Terms borrowed from the physical sciences, like "cause" (Haldane's preferred word), "factor" (Wright's preferred word), and finally "mechanism" (Dobzhansky's and Huxley's preferred word), slowly supplanted the term and view of selection as agent, although Huxley viewed selection as both agent and mechanism simultaneously and Fisher still viewed it as an agent (see discussion below).[78]

All three mathematical modelers were attempting to bring biology up to par with the physical sciences, as they drew on, and modeled after, the repertoire of the physical sciences.[79] This was the case for Haldane, whose major work was entitled *The Causes of Evolution* and who was a supporter of the Unity of Science Movement,[80] and to a lesser extent for Sewall Wright, for whom natural selection was a mathematical "factor."[81] The wish to bring biology to heel within the physical sciences was most intense in Fisher, who clearly stated his intent to model evolution after physics and chemistry. The title of Fisher's 1930 book, *The Genetical*

[78] These terminological variations are still present in evolutionary biologists' vocabularies, though "mechanism" is the preferred word.

[79] I am not here claiming that the modelers were strict physicalists who wished to reduce biology to physics, but that they were drawing from exemplars in the more exact and rigorous physical sciences. The exemplar of *the* scientific method was Newtonian physics, as articulated through Newton's Rules of Reasoning and then transmitted to a wider audience—including Darwin—through philosophers like William Whewell and John Herschel. Fisher, who espoused a form of indeterminism, and especially Wright, who upheld a form of panpsychic dualism (or panpsychism), could hardly be deemed strict physicalists. For a discussion of the metaphysics of Fisher and Wright, see M. J. S. Hodge, "Biology and Philosophy (Including Ideology): A Study of Fisher and Wright," in Sahotra Sarkar, ed., *The Founders of Evolutionary Genetics* (Dordrecht: Kluwer, 1992), pp. 231–93.

[80] In 1936 Haldane was present at the Second International Congress for the Unity of Science held in Copenhagen. The title of his paper was "Analysis of Causality in Genetics" (Box 3, Folder 2, Joseph Regenstein Library, University of Chicago).

[81] The mathematical underpinning of Wright's conception of evolution is indicated by his notion of natural selection as a factor. Interestingly, though his evolutionary work was equally significant, Sewall Wright never published a major book in the 1930s comparable to those of Fisher and Haldane. For both Fisher and Haldane, who had greater political ambitions, a book-length treatment of evolution made more sense. Wright eventually published a lengthy multivolume treatise on evolution between the years 1968 and 1978: *Evolution and the Genetics of Populations*, 4 vols. (Chicago: University of Chicago Press, 1968–78).

Theory of Natural Selection, raised the status of natural selection from fact to theory, as Fisher articulated his "fundamental theorem of natural selection."[82]

As natural selection became measurable and testable, and came to be seen as a causo-mechanical explanation for organic change, much of the metaphysical and speculative status of the phenomenon was removed.[83] As natural selection and the other "factors" in evolution came increasingly to be seen in physicalist terms, the factors took on a causo-mechanical reality. Simultaneously, these "mechanisms"—now made to be mechanisms—came to be aligned with the material basis of evolution. By the mid-1930s all the components had been assembled that could make evolution as mechanistic and materialistic a science as possible, grounded firmly in the Hardy–Weinberg Equilibrium Principle, and following "observation and experiment," but in a manner that also would permit the independence of the biological sciences. The bringing together of the mechanical cause of evolutionary change with the material basis of evolution would prove to be the most difficult, but key, feature in the making of a science—now at least rivaling and arguably analogous to that of Newtonian physics—of evolution. This was to come through the work of Sewall Wright, but especially through the lifework of Theodosius Dobzhansky.

A keen, self-taught naturalist, early aware of the variational features in natural populations of organisms, Dobzhansky began his career as systematist, studying classical problems of classification.[84] Unlike many of

[82] Fisher's fundamental theorem was meant explicitly to resemble the law that held the "supreme position among all laws of nature," the second law of thermodynamics. Among other resemblances, he noted that both had the properties of populations or groups of aggregates, and that both were statistical laws. If the chemists could have such a supreme law, Fisher argued, the biological sciences could have one as well. Natural selection, as Fisher described it, became the biological analogue of the second law of thermodynamics.

[83] Though many metaphysical features of selection were removed, I will argue shortly that enough of the teleology was left behind for there to be a purposive and independent science of life.

[84] Dobzhansky had also worked with Iurii Filipchenko in Russia. For a historical discussion on the life and thought of Dobzhansky with special attention to the Russian origins of his thought, see Mark B. Adams, ed., *The Evolution of Theodosius Dobzhansky* (Princeton: Princeton University Press, 1994). See also idem, "The Founding of Population Genetics: Contributions of the Chetverikov School, 1924–1934," *Journal of the History of Biology* 1 (1968): 23–39; and idem, "Towards a Synthesis: Population Concepts in Russian Evolutionary Thought, 1925–1935," *Journal of the History of Biology* 3 (1970): 107–29. For an intellectual biography of Dobzhansky, see W. B. Provine, "Origins of the Genetics of Natural Populations Series," in R. C. Lewontin et al., eds., *Dobzhansky's Genetics of Natural Populations I–XLIII* (New York: Columbia University Press, 1981), pp. 5–83. See also Howard Levene, Lee Ehrman, and Rollin Richmond, "Theodosius Dobzhansky up to Now," in Max K. Hecht and William C. Steere, eds., *Essays in Evolution and Genetics in Honor of Theodosius Dobzhansky: A Supplement to Evolutionary Biology* (New York: Appleton-

his more typological Anglo-American counterparts in systematics, and heir to a Russian population approach to genetics and systematics, Dobzhansky had early on developed a dynamic, evolutionary approach to systematics that stressed the geographic variation of natural populations of organisms he studied. This natural emphasis on populational thinking, the study of variation in different geographic regions, an early appreciation for genetics, and his training in Russia gave Dobzhansky an integrative edge that allowed him to see the possibility of both populational and genetical evolutionary points of view. Leaving his Russian mentors (and his Russian origins), Dobzhansky immigrated to the United States, where he sought to learn the latest in "transmission" or "classical" genetics associated with Thomas Hunt Morgan and his "group" in the "fly room" at Columbia. This training subsequently prepared him to synthesize diverse branches of biological knowledge within a coherent, evolutionary worldview.

For Dobzhansky, who apprenticed in the laboratory of classical geneticist Morgan, the material basis for evolution—that which Darwin and his heirs had sought—became the gene, arrayed in linear fashion on the material carriers of heredity, the chromosomes.[85] Close examination of the behavior of salivary chromosomes in natural populations of *Drosophila pseudoobscura* led to the determination of inversion frequencies and formed the basis of what is known as Dobzhansky's studies of the genetics of natural populations.[86] Genetic mechanisms that accounted for microevolutionary change (evolution below the species level) also accounted for larger-scale or macroevolutionary change (evolution includ-

---

Century-Crofts, 1970), pp. 1–41. Dobzhansky actually learned much of his classical genetics through dialogue with A. H. Sturtevant. Much of his early work with the Morgan group had been on the preferred laboratory organism of geneticists, the fruit fly, *Drosophila melanogaster*.

[85] The chromosomal theory of heredity, sometimes referred to as the Sutton–Boveri theory, had been articulated in 1902–3. It pointed to the chromosomes as the material carriers of heredity. The "gene"—as constructed by workers at the turn of the century to become the unit or particle of heredity—was seen to be carried on the material of the chromosomes, made observable by the advent of dyes, stains, sectioning techniques, and the imaging technology of the microscope. The gene can be viewed as an analogue of the particle in Newtonian physics; hence, the debates focusing on the gene bear some resemblance to the early debates in Newtonian physics. The construction of the gene culminated with the work of T. H. Morgan and his group and the publication of *The Theory of the Gene* in 1926: T. H. Morgan, *The Theory of the Gene* (New Haven: Yale University Press, 1926). This followed on the foundational work summarized in T. H. Morgan, A. H. Sturtevant, H. J. Muller, and C. B. Bridges, *The Mechanism of Mendelian Heredity* (New York: Henry Holt, 1915).

[86] For a historical description of the "GNP" series, see Provine, "Origins of the Genetics of Natural Populations Series."

ing, and above, the species level); this continuum therefore accounted for mechanisms of speciation that had a genetic grounding.[87]

The alignment of the material basis with the mechanical cause of evolution bore close resemblance to Newtonian mechanics. The gene, which after the work of Morgan and his group became an entity that functioned as the particle of heredity,[88] became the unit of evolutionary change; selection would become the primary driving motive force to propel evolutionary change.[89] The gene—by the Morgan school's standards—had been constructed to particularize and individuate, and at the same time to limit the rate of change, and mutations became the determinants or analogues to fluxions of evolutionary change.[90] Phenotypic saltations, observable as the result of the chromosomal alterations prevalent in model plant organisms like *Oenothera lamarckiana*, were therefore to be tempered through the adoption of model animal systems like *Drosophila melanogaster*, which would restrict evolutionary change to the level of small, individual differences. Such point mutations instead of macromutations would thus limit the rate of change. The final result, Dobzhansky's "synthesis," offered an account of evolutionary change that would limit—and make deterministic—the rate of evolutionary change.[91] From then on, measures would be taken to calculate and determine evolutionary rates of change. Evolution, in turn, could be redefined as the "change in gene frequencies." Viewed as a problem in accounting for change, the Hardy–Weinberg Equilibrium Principle, which effectively set the conditions under which there would be no evolutionary change, converted the variables of natural selection, mutation, population structure, random genetic drift, migration, and systems of mating into causal explanations for evolutionary change. As in Newton's laws of motion, there could be no evolutionary change without one of these causes of evolutionary change. Evolutionary change would thus be constructed on

[87] Dobzhansky introduced notions of "reproductive isolating mechanisms" to account for the origin of species.

[88] Particularistic theories of heredity had been favored over blending theories for this reason.

[89] This substituted selection pressure for the mutation pressure present in the mutation theory of Hugo de Vries. Dobzhansky's evolutionary genetics therefore stressed the *creative* element of natural selection, whereas de Vries had stressed the *eliminative* element. For Dobzhansky, the unit of evolutionary change was the gene; for de Vries, it was the species. Debates subsequently took place over the exact target of selection.

[90] The establishment of particulate theories of heredity effectively dispelled belief in blending theories. This had been one of the key contributions that the Mendelians had made to the synthesis.

[91] Thus, though Dobzhansky himself studied large-scale chromosomal arrangements like inversions and translocations, the actual result of his "synthesis" would be to limit the rate of change.

models of physical change so that evolution would demonstrate law-like regularities analogous to the law-like regularities in Newtonian physics.

Dobzhansky had drawn heavily, consciously, on the "classical" genetics of the Morgan school, which in its mechanistic and materialistic nature most closely resembled "classical" physics. Genetics (and the physical world of the gene) was used as the grounding for Dobzhansky's new "evolutionary genetics" (a new phrase) and formed the basis for his belief in the continuum between microevolution and macroevolution (which stretched from the gene, to the human, to human culture).[92] The title of his book published in 1937 reveals this grounding: *Genetics and the Origin of Species* offered a framework that brought together the material basis for evolution, determined first through the work of geneticists, with causo-mechanical explanations—made mechanical through the models—for evolution.[93] At the same time, Dobzhansky offered mechanisms for speciation, or in his terms, "isolating mechanisms," to account for the geographic origin of species in natural populations. Hence, the "evolutionary synthesis," held by historical commentators to involve the synthesis between "genetics and selection theory," can be reinterpreted as the bringing together of the material basis of evolution (the gene) with the mechanical cause of evolutionary change (natural selection) to make a mechanistic and materialistic science of evolution that could rival Newtonian physics. Dobzhansky's alignment of the material basis with the mechanical cause of evolution in turn gave rise to a science obeying the methodology of "observation and experiment," evolutionary genetics, which had the most important effect of working in natural populations of *Drosophila* (in a "Back to Nature" move). Dobzhansky was aware of this grounding and reflected on it in his oral memoirs of 1962:

> Genetics is the first biological science which got in the position in which physics has been in for many years. One can justifiably speak about such a thing as theoretical mathematical genetics, and experi-

[92] Dobzhansky drew on Chetverikov's terms "microevolution" and "macroevolution." Julian Huxley had also promoted the phrase "evolutionary genetics."

[93] Theodosius Dobzhansky, *Genetics and the Origin of Species* (New York: Columbia University Press, 1937). The title was actually given to him by Columbia geneticist L. C. Dunn when he invited Dobzhansky to revive the Morris K. Jesup Lectures in 1936. Though the title that Dunn suggested was ambitious, Thomas Hunt Morgan, Dobzhansky's supporter in Pasadena, thought that such a grand project would encourage Dobzhansky to bring out his best. Thus the lectures directly provided incentive for Dobzhansky to address the subject. For the detailed story leading up to the lectures and publication of Dobzhansky's book, see William B. Provine, "Origin of Dobzhansky's *Genetics and the Origin of Species*," in Mark Adams, ed., *The Evolution of Theodosius Dobzhansky* (Princeton: Princeton University Press, 1994), pp. 99–114. See also the discussion below.

mental genetics, just as in physics. There are some mathematical geniuses who work out what to an ordinary person seems a fantastic kind of theory. This fantastic kind of theory nevertheless leads to experimentally verifiable prediction, which an experimental physicist has to test the validity of. Since the times of Wright, Haldane, and Fisher, evolutionary genetics has been in a similar position.[94]

Importantly, this mechanistic and materialistic framework grounded in genetics, and ultimately in the Hardy–Weinberg Equilibrium Principle, would also account for higher-level phenomena, including not only the origin of species, but also the origin of humans, the mind, and culture, now unifiable and reducible to lower-level phenomena. The continuum, as it was constructed, could therefore legitimate as it connected inquiry into the mechanics of change in human culture with the material on which the mechanics acted, as well as introducing inquiry into a mechanistic and materialistic theory of mind—all ultimately reducible to the gene. The continuum that Dobzhansky's framework provided would also ground—through a fundamental logic—the social sciences.[95] But for Dobzhansky and other evolutionists, this continuum involved only partial reduction.[96] While the possibility of reduction from higher levels to lower levels existed, measures could and would be taken to ensure that certain phenomena were not subject to reduction to the physical world. Emergent properties, which in some measure could be considered meta-

[94] As cited on p. 277 of Provine, *Sewall Wright*, from the Oral Memoir of 1962, pp. 500–501, Columbia University Archives, New York. Celebrating the fifty-year "Jubilee of Genetics," Julian Huxley echoed Dobzhansky: "In the 50 years since Mendel's Laws were so dramatically rediscovered, genetics has been transformed from a groping incertitude to a rigorous and many-sided discipline, the only branch of biology in which induction and deduction, theory and experiment, observation and comparison have come to interlock, in the same sort of way that they have for many years done in physics" (Julian Huxley, "Genetics, Evolution and Human Destiny," in L. C. Dunn, ed., *Genetics in the 20th Century: Essays on the Progress of Genetics During the First 50 Years* [New York: Macmillan, 1951], p. 591).

[95] Recall the arrangement of the sciences represented in fig. 1. With the maturation and institutionalization of the social sciences in the 1960s, the nature versus nurture debates would sweep across universities in the United States and Britain. For historical background, see Hamilton Cravens, *The Triumph of Evolution: American Scientists and the Heredity–Environment Controversy, 1900–1941* (Philadelphia: University of Pennsylvania Press, 1978; repr. as *The Triumph of Evolution: The Heredity–Environment Controversy*, 1900–1941 [Baltimore: Johns Hopkins University Press, 1988]).

[96] See fig. 1 for the arguments for "Emergence" in the sciences through "legitimate materialisms." Ernst Mayr was to support a partially emergentist model; see his discussion in *The Growth of Biological Thought*, pp. 63–67. Social scientists also adopted emergentist models to support nurture over nature. Others, like William B. Provine, took an extreme reductionist model to its logical conclusion and denied the possibility of free will.

physical,[97] would therefore be evoked by these biologists to make room for the possibility of a meaningful life, devoid of complete determinism; at the same time, these properties would make possible the independence of the biological sciences.

This continuum was the most important undergirding feature of Dobzhansky's framework. So long as the continuum between the gene and the human (grounded in Newtonian physics and chemistry and the Hardy—Weinberg Equilibrium Principle) existed, there would be unity between the levels of evolution. Though the ultimate grounding was to be the physical world of the gene, for Dobzhansky (and others) there would be enough of the metaphysical world left behind at higher levels (like mind and culture) to make possible a meaningful life, avoiding a completely genetically deterministic view.[98] Reducible to the physical world and ultimately to a mathematical logic, Dobzhansky's continuum would function—later—as a unifying argument and a unification event within positivistic standards.[99] Obeying the common method of observation and experiment, reducible through logic, and obeying axiomatic mathematical principles, evolution and biology would in turn bind and unify the sciences.

The quantification and measurability of these factors, the consequent quantification of evolution, the alignment of the mechanical cause with the material basis for evolution, and the beginnings of the adoption of natural selection as the primary mechanism of evolution led simultaneously to the ejection of "alternative" mechanisms of evolution.[100]

[97] Whether emergent properties can be deemed metaphysical is a contentious issue for philosophers of biology. Though I would agree that there are fundamental distinctions between these terms, all function in the same way in a disciplinary sense. Emergentism functions in the same manner as vitalism, teleology, and other unarticulated metaphysical elements: all "lift" biology from complete reduction to the physical sciences.

[98] As a group (comprising the discipline) the architects of the evolutionary "synthesis" would negotiate and strike just the right balance between mechanistic materialism and some form of emergentism (see the discussion below). Those who rejected a completely mechanistic and materialistic framework include E. Mayr, R. A. Fisher, S. Wright, C. H. Waddington, D. Lack, B. Rensch, and J. Huxley, as well as Dobzhansky. Of the group of "synthesis" architects, the only person who upheld mechanistic and materialistic evolution was Simpson, for whom evolution was a historical process, one that dealt with unique historical events. In Simpson's historical perspective, chance events and contingencies introduced indeterminism into the evolutionary system. Geneticists like T. H. Morgan, A. H. Sturtevant, and C. Bridges upheld mechanistic and materialistic frameworks. See the discussion below on R. Goldschmidt. See also the discussion below on "Biological Autonomy in the Post-*Sputnik* Biological Sciences."

[99] In his historical work, Ernst Mayr has explicitly stated that a unification event took place during the synthesis; see "On the Evolutionary Synthesis and After," chap. in *Toward a New Philosophy of Biology: Observations of an Evolutionist* (Cambridge, Mass.: Belknap, 1988), pp. 525–54. William B. Provine has also echoed this belief.

[100] They became "alternative" once biologists could make one of them primary.

Directed evolution (aristogenesis, nomogenesis, orthogenesis, etc.), Lamarckian inheritance, and emergent evolution, among others, were ejected from mainstream evolutionary studies as what appeared to be a narrowing or streamlining of acceptable evolutionary theory took place. This is the phenomenon that Provine has characterized as the "evolutionary constriction."[101] Historian John Greene has similarly argued that these same "metaphysical" elements were removed with the modern synthesis.[102] With these metaphysical elements ejected, a unification event— based more on a Machian unification argument—was beginning to take place at this time.

But one complication was present in aligning the material basis of evolution with the mechanical cause of evolutionary change. For Dobzhansky as for Wright, strongly selectionist models of evolution did not resonate with their initial view of evolution in natural populations.[103] Strongly selectionist models had been favored by R. A. Fisher, the same individual who persisted in viewing natural selection as an agent. For Fisher, the belief in the agency and power of selection was inextricably linked to his deeply held eugenical commitments.[104] His 1930 book devoted a great deal of space to a discussion of the eugenicist agenda. If selection had enough agency (and at the same time were a mechanical principle), then adaptability and "improvement" of humans would be all

[101] William B. Provine, "Progress in Evolution and Meaning in Life," in Matthew Nitecki, ed., *Evolutionary Progress* (Chicago: University of Chicago Press, 1988), pp. 49–74.

[102] See John C. Greene's essay, "The History of Ideas Revisited," *Revue de synthèse* 4 (1986): 201–27. Provine knew that these alternative theories were ejected in "the evolutionary constriction" because books on evolutionary topics appeared to be getting smaller in size as they left these topics out. Greene states explicitly that theories that smacked of finalism were considered "metaphysical" by biologists. For Greene, the removal of these was part of the synthesis. Neither had explicitly recognized that this could constitute a unificatory event within a Machian positivistic framework, though Greene has explicitly recognized the larger positivistic currents in the modern synthesis.

[103] Many naturalists had opposed the view that natural selection was responsible for interspecific and interpopulational variation. Among the most influential in making this argument were G. C. Robson and O. W. Richards, *The Variation of Animals in Nature* (London: Longmans, 1936); and G. C. Robson, *The Species Problem* (London: Oliver and Boyd, 1928). Others included J. T. Gulick, H. E. Crampton, and Frances Sumner. These individuals thought that interspecific and interpopulational differences were nonadaptive and in some measure arose from chance events. William B. Provine discusses the work of these naturalists in his biography of Sewall Wright. This feature of the Wright and Dobzhansky framework is also discussed at length by Provine in his biography of Wright.

[104] For a detailed discussion of Fisher, the success of selection, and the Oxford school of ecological genetics, see John R. G. Turner, "Random Genetic Drift, R. A. Fisher, and the Oxford School of Ecological Genetics," in Lorenz Krüger, Gerd Gigerenzer, and Mary S. Morgan, eds., *The Probabilistic Revolution* (Cambridge: MIT Press, 1987), vol. 2, pp. 313–54. See also the discussion of Fisher in Mary M. Bartley, "Conflicts in Human Progress: Sexual Selection and the Fisherian 'Runaway,'" *British Journal for the History of Science* 27 (1994): 177–96.

the more possible and rapid. Just as important as the commitment to human "improvement" was the closely related "progressive" view of evolution, made all the more obvious with the publication of Huxley's *Evolution: The Modern Synthesis*. Wright and Dobzhansky's initially strong support of random genetic drift diminished as the result of a combination of factors: the drive for the improvement of humans and the increasing necessity for progressive evolution within a positivistic theory of knowledge.[105] In such a philosophical framework, more strongly selectionist models would be favored by biologists who patterned themselves after physicists at the same time that they pointed the way to the improvement of humanity, allowed for adaptability to a rapidly changing world, and thus painted a progressive and optimistic picture of the same world.[106] This was the packaging that Julian Huxley was to put together; and Huxley's packaging was to prove the most immediately efficacious—to a wide audience—to lend both unity and independence to the biological sciences. The ultimate push to adopt more selectionist and adaptationist models would come from outside the local network around Dobzhansky and Wright that wished to preserve some measure of progress. Evolutionary models favoring random genetic drift, which enforced a stochastic view of evolution—and culture—would not be favored in a postwar frame of mind seeking to improve the world.[107] So powerful would be the felt need for a progressive, selectionist, and adaptationist framework that in the 1940s even Dobzhansky and Wright would come to adopt more strongly selectionist models.

For the time being, however, the alignment of the mechanical cause (selection) and the material basis (the gene) for evolution, and the ejection of enough of the metaphysical components, along with the establishment and extension of experimentation in natural populations of organisms began to legitimate the long-beleaguered evolutionary studies and unify the fractured biological sciences within the continuum from

---

[105] Discussion on progressive evolution follows below.

[106] For a discussion on Wright and drift, see William B. Provine, "The Development of Wright's Theory of Evolution: Systematics, Adaptation and Drift," in Marjorie Grene, ed., *Dimensions of Darwinism*, pp. 43–70. For a detailed discussion on reasons for Dobzhansky's shift, see John Beatty, "Dobzhansky and Drift: Facts, Values, and Chance in Evolutionary Biology," in *The Probabilistic Revolution*, pp. 271–311. Beatty argues that Dobzhansky shifted his favoring of genetic drift to his "balance" position in opposition to Muller, who favored an opposing "classical" position.

[107] Too much stochasticity in the form of random genetic drift made the system too unpredictable. A middle ground—deterministic enough to make predictions, but having enough indeterminism—through some sort of metaphysical or emergentist phenomena would be favored. At the same time, this balance made possible a meaningful life with humans as agents of their own free will.

the gene to the human. Purged of its unacceptable metaphysical elements and grounded in the mathematics of the Hardy–Weinberg Equilibrium Principle, evolution was becoming a purified science more secure in its foundations, a science that could begin to meet even some of the earlier criticisms raised by Woodger and others.[108] Evolution would increasingly be used to lend both unity and autonomy to the biological sciences. Evolution in turn would be used not only to unify biology, but also the rest of the sciences. In the grand sweep of the history of science, the "modern" synthesis of evolution between the material of the gene and the mechanism of natural selection within the continuum from the gene to the human would be a moment when what was thought to be "meta"-physical would "touch" the ground: the "ultimate" questions of the meaning of life and human origins would be cojoined and reducible through logic to the physicalist, mechanistic, and materialistic frameworks of the physical sciences. Taking this profound message to a wider popular audience in his 1949 book *The Meaning of Evolution*, G. G. Simpson would write:

> It is assumed that a material universe exists and that it corresponds with our perceptions of it. The existence of absolute, objective truth is taken for granted as well as the approximation to this truth of the results of repeated observations and experiments. That such assumptions are debatable is evident from the violence with which they have been debated at various times. In practice, however, we all have to take it either that they are true or that we necessarily proceed *as if* they were true. Otherwise there is no meaning in science or in any knowledge, or in life itself, and no reason to enquire for such meaning.[109]

BINDING THE HETEROGENEOUS PRACTICES OF BIOLOGY: DOBZHANSKY'S *Genetics and the Origin of Species* AND THE COLUMBIA BIOLOGICAL SERIES

The genetic grounding Dobzhansky had adopted rapidly became the foundation for what Woodger and others had pointed out were the unstable biological sciences. What for Woodger had appeared to be a ground of "fundamental biological antithesis," difficult to harmonize and synthesize because of underlying contradictions, became the common ground of genetics and selection theory—the "synthesis" that Dob-

---

[108] The sense that evolution is one of the "softer" biological sciences is still prevalent.
[109] George Gaylord Simpson, *The Meaning of Evolution: A Study of the History of Life and of Its Significance for Man* (New Haven: Yale University Press, 1949), p. 7.

zhansky's evolutionary genetics had achieved. Dobzhansky's genetics thus became the ground on which the heterogeneous practices of the biological sciences were stabilized and bound.

A group of biologists, in close dialogue with the extraordinarily charismatic Dobzhansky, began to link up and legitimate their *own* practical concerns with the framework provided by Dobzhansky's "synthetic" evolutionary genetics.[110] By 1936 Dobzhansky was preparing to present his evolutionary genetics—linked with these practical concerns—to an even wider audience of evolutionists by converting his Columbia University Jesup Lectures of 1936 into what became the first textbook of evolution.[111] Dobzhansky's evolutionary genetics thus became codified and disseminated with the publication in 1937 of *Genetics and the Origin of Species*, which served to initiate practitioners into the new evolutionary craft and take the new practices from Dobzhansky's local network to the wider evolutionary audience. But Dobzhansky's book, which drew heavily on calculations of gene frequencies and defined evolution in terms of changes in those frequencies, was hardly about evolution as a whole.[112] Nor could the discipline of genetics by itself account for the entirety

[110] See Clifford Geertz's reflections on charismatic figures in *Local Knowledge* (New York: Basic, 1983); and see also idem, *The Interpretation of Cultures* (New York: Basic, 1973). The authors of the Columbia Biological Series had started engaging in dialogue (both published and personal) with one another in the mid-1930s. Before these books appeared all the authors had published series of papers on related themes, which the others had been reading. Hence Dobzhansky's framework developed as a result of a multidirectional traffic of influence, negotiated and renegotiated by members inside and outside Dobzhansky's local group. Group dynamics were complex. Dobzhansky had personally drawn in Stebbins and others to his evolutionary genetics. Dobzhansky and Huxley had been in close touch with each other and had made moves to start an official society in 1939, only to be thwarted by wartime preparations. Mayr was also in contact with Huxley and read and commented on early chapters of Huxley's 1942 book.

[111] The Jesup Lectures had not been given since 1910. The topic of evolution had been especially suitable because the first problem of the original series had been evolution as a problem of history as expressed by H. F. Osborn in his *From the Greeks to Darwin*. Three editions of Dobzhansky's influential book were issued: 1937, 1941, 1951. The short interval of time between each successive edition attests to the increasing level of activity in evolutionary studies during these years. *Genetics and the Origin of Species* was considered essential reading for all evolutionists at least until the 1970s. See Howard Levene, Lee Ehrman, and Rollin Richmond, "Theodosius Dobzhansky up to Now," for their discussion of how Dobzhansky's previous work was assimilated for the Jesup Lectures and the influence of *Genetics and the Origin of Species*. See also William B. Provine's contribution to Mark Adams, ed., *The Evolution of Theodosius Dobzhansky*, "The Origin of Dobzhansky's *Genetics and the Origin of Species*," pp. 99–114.

[112] Ernst Mayr was to play an increasingly vital role in pointing out that this definition of evolution, which did not account for the origin of discontinuities (especially speciation), was incomplete. In the late 1950s he was to describe the narrow population geneticists' view of evolution as "bean-bag genetics," and to promote the transformational features of evolution. See the discussion below entitled "Biological Autonomy in the Post-*Sputnik* Biological Sciences."

of the evolutionary process that appeared to be much more complex and called for contributions from other biological sciences. If genetics indeed had a place in accounting for the origin of species, as Dobzhansky argued, then so too did other closely neighboring biological disciplines.

The audience for Dobzhansky's book consisted of diverse groups within evolutionary studies who read, responded to, and further legitimated and extended the evolutionary framework provided by Dobzhansky's evolutionary genetics. In so doing they began to bind the heterogeneous practices of evolution into an evolutionary network grounded in genetics and selection theory for a wide audience of interested biologists. The publication of *Genetics and the Origin of Species* signaled and served as catalyst for the publication of a series of books called the Columbia Biological Series: Ernst Mayr's *Systematics and the Origin of Species*, G. G. Simpson's *Tempo and Mode in Evolution*, and G. L. Stebbins' *Variation and Evolution in Plants*.[113] These books were written by individuals who, engaging in dialogue with Dobzhansky, in turn legitimated as they grounded *their* disciplines with Dobzhansky's experimental evolutionary genetics. Each of the authors had inherited from his discipline a different set of problems, practical in nature, which Dobzhansky's book and its evolutionary program offered in some measure to resolve. Each read his own meaning into Dobzhansky and responded in turn. And each offered an amendment to Dobzhansky's framework.

To Mayr, the appeal of Dobzhansky's framework lay in its populational features and its support of a biological species concept.[114] For practicing ornithologists like Mayr in the 1930s and 1940s, successive populational samples instead of the solitary "type specimen" had become the working unit of the taxonomist. Dobzhansky's framework with its emphasis on natural populations and subspecies thus made tractable the problems of working taxonomists, and gave plausible causo-mechanical explanations, or mechanisms for speciation (like Dobzhansky's "isolating mechanisms"), which took into account geographic variation within slow, gradual rates of change. At the same time Mayr's stress on geographic variation within slow, gradual rates of change would also effec-

---

[113] Ernst Mayr, *Systematics and the Origin of Species* (New York: Columbia University Press, 1942); George Gaylord Simpson, *Tempo and Mode in Evolution* (New York: Columbia University Press, 1944); G. Ledyard Stebbins, Jr., *Variation and Evolution in Plants* (New York: Columbia University Press, 1950).

[114] It will be recalled that Dobzhansky's interest in populations was an outcome of his background in systematics, which he had inherited from his Russian mentors. See the relevant literature on Dobzhansky noted above.

tively counteract the argument for saltationist evolution that had gained currency with the recent appearance of Richard Goldschmidt's 1940 book, *The Material Basis of Evolution*.[115] Dobzhansky's framework also had the most pleasing aspect of returning systematics to the field. But while Dobzhansky emphasized the populational features of evolution and opened inquiry into the mechanisms of speciation, his book failed to discuss in sufficient detail the topic heralded by the title: the origin of species.[116] In stressing a definition of evolution based solely on gene frequencies, Dobzhansky's framework did not sufficiently take into account the primary concern for systematists and a central component of evolution: accounting for the origin of organic discontinuities. So, too, Dobzhansky's scheme had stressed too heavily genes as the targets of selection but had not addressed selection at the level of the individual organism, the domain of evolution of interest to population-inclined naturalist-systematists. Redressing a perceived imbalance in Dobzhansky's emphasis on genetics to the exclusion of systematics, Mayr's *Systematics and the Origin of Species*—as the title indicates—was meant to be a direct response to *Genetics and the Origin of Species*.[117]

To Simpson, who had especially endorsed the quantification of evolution,[118] the measurability of natural selection as outlined in Dobzhansky's book meant that paleontology could be rid of the metaphysical horrors of directed evolution which, following H. F. Osborn's directives, had pervaded the discipline. In Simpson's view, natural selection, modified in such a manner that "quantum evolution" could take place to account for accelerations in evolutionary change, made possible the resolution of problems faced by practicing paleontologists. The paleontological framework of Simpson's *Tempo and Mode in Evolution*, in turn, also had a pow-

[115] Ernst Mayr writes that "Goldschmidt's 1940 book was a temporary setback; this continuing resistance induced me to to devote twenty-four pages of my 1942 book to a series of proofs of geographic speciation. By permitting slow evolution, the gradual acquisition of isolating mechanisms, and the entering of new niches or adaptive zones, geographic speciation is ideally suited for an application of the neo-Darwinian interpretation of evolution" (Ernst Mayr, "The Role of Systematics in the Evolutionary Synthesis," in Ernst Mayr and William B. Provine, eds., *The Evolutionary Synthesis*, p. 131).

[116] Ironically, Dobzhansky's book did not actually include even a chapter-length substantive discussion on speciation.

[117] Ernst Mayr's own historical sense, correctly so, is that the synthesis was in part a switch from a typological to a populational way of thinking. It was—for him. See Mayr's "Prologue" to Ernst Mayr and William B. Provine, eds., *The Evolutionary Synthesis*.

[118] For a discussion of Simpson's endorsement, and his own abilities in quantitative methods, especially statistics, see Léo F. Laporte, "Simpson on Species," *Journal of the History of Biology* 27 (1994): 141–59. Laporte also discusses the influence of Anne Roe Simpson on the development of Simpson's application of quantitative methods in paleontology.

erful effect in validating Dobzhansky's framework.[119] By providing observable evidence of the evolutionary process, Simpson was to legitimate evolution as a historical science. The "woefully inadequate" fossil record was used once again to buttress an evolutionary perspective favoring slow, gradual change with enough room for evolutionary quantum "jumps."[120]

For Stebbins, the appeal of Dobzhansky's framework was in making workable (within an evolutionary framework) the chaotic profusion of plant data that had accumulated over the years. By the 1940s the work of practical breeders, agriculturists, and horticulturists had made observable the curious behaviors of chromosomes and the noncharacterizable mating habits of plants. These problematic phenomena became tractable once Stebbins envisioned them as genetic systems (a notion he borrowed from cytogeneticist C. D. Darlington)[121]—apoximis, hybridization, and polyploidy—which themselves were subject to the mechanism of selection. To plant taxonomists, moreover, the biological species concept offered by Dobzhansky was one way of making tractable the long-held problem of species in plants. With open or indeterminate genetic systems, plants also had complex variation patterns, since genotypic and phenotypic responses were difficult to distinguish; as a result, a belief in Lamarckian or "soft" inheritance had been widespread in botanical circles. Natural selection, as a mechanism, helped dispel such speculative points of view and rid botany of the belief in Lamarckian inheritance. So,

[119] George Gaylord Simpson, *Tempo and Mode in Evolution* (New York: Columbia University Press, 1944); this was subsequently rewritten and updated as *The Major Features of Evolution* (New York: Columbia University Press, 1953).

[120] Stephen Jay Gould's historical sense that for paleontologists the synthesis led to the rejection of directed evolution is also correct—for practicing paleontologists. See Stephen Jay Gould, "G. G. Simpson, Paleontology, and the Modern Synthesis," in Ernst Mayr and William B. Provine, eds., *The Evolutionary Synthesis*, pp. 153–72. Bernhard Rensch's *Evolution above the Species Level* played a similar role. See Bernhard Rensch, *Evolution above the Species Level* (New York: Columbia University Press, 1959). The first German edition appeared in 1947; the Columbia edition was the English translation of the second German edition that appeared in 1954.

[121] Stebbins drew heavily on C. D. Darlington's *Recent Advances in Cytology* (Philadelphia: Blakiston's, 1932). Darlington subsequently rewrote his final chapter on genetic systems, and published it as *The Evolution of Genetic Systems* (Cambridge: Cambridge University Press, 1939). Darlington's work on genetic systems had been extremely controversial at the time. For an engaging account of the controversy that *Recent Advances in Cytology* precipitated, see Hampton L. Carson, "Cytogenetics and the Neo-Darwinian Synthesis," in Ernst Mayr and William B. Provine, eds., *The Evolutionary Synthesis*. Carson writes: "The book was considered to be dangerous, in fact poisonous, for the minds of graduate students. It was made clear to us that only after we had become seasoned veterans could we hope to succeed in separating the good (if there was any) from the bad in Darlington. Those of us who had copies kept them in a drawer rather than on the tops of the our desks" (p. 91).

too would the mechanism of natural selection help rid botany of the mutationism long a holdover from the early days of *Oenothera* genetics. For Stebbins, therefore, the dialogue with Dobzhansky led to his account of *Variation and Evolution in Plants*.[122]

With the exception of the disciplines of embryology and physiology, for which the gene could not be reconciled with mechanical embryological or physiological principles,[123] the Columbia Biological Series bound the heterogeneous assemblage of practices of closely neighboring disciplines of the biological sciences. These books in systematics, paleontology, and botany represented only a microcosm,[124] however, and an American microcosm at that,[125] of the work in evolutionary studies drawing on the mechanistic and materialistic frameworks that had been established by the mathematical modelers in conjunction with field biologists

[122] There is much to Mayr and Provine's historical sense that botany was in some manner delayed in "entering the synthesis." Botany by the late 1940s, unlike systematics and paleontology, consisted of a much more heterogeneous assemblage of practices: taxonomy, morphology, genetics, ecology, paleobotany, and the like. Stebbins, as a practitioner in all these areas, had to bring into line a great deal more data: at 643 pages, his book was the longest and last of the synthesis classics. See Smocovitis, "Botany and the Evolutionary Synthesis."

[123] These two disciplines had stressed the transformational instead of the populational features of evolution. Both had been buffeted by the extremes of vitalism and mechanism so that a middle ground would be especially difficult to attain. That embryology was "left out" of the synthesis has been a subject for discussion by historians of the synthesis period. Proponents of synthesizing embryology with evolution and genetics included Gavin de Beer, Richard Goldschmidt, and Conrad H. Waddington. De Beer's *Embryology and Evolution* (Oxford: Clarendon, 1930) helped dispel the Lamarckism that had accompanied recapitulation theory. See the discussion on Goldschmidt below. Conrad Waddington increasingly opposed the view that a synthesis between biological disciplines had actually occurred. So, too, did many experimental biologists. See, for instance, the volume entitled *Evolution* from The Symposium of the Society for Experimental Biology (New York: Academic Press, 1953). The volume contains the papers read at an Oxford Symposium of 1952. J .B. S. Haldane admitted in the preface that the synthesis was by no means complete. Drawing on a developmental metaphor, he claimed it was in a current "instar" defined by books such as those of Huxley, Simpson, Dobzhansky, Mayr, and Stebbins. This was the volume that also published C. H. Waddington's famous opposition to the synthesis, *Epigenetics and Evolution* (pp. 186–99). Opponents of the legitimacy of evolution as science, especially many embryologists and physiologists, in the presynthesis period never really accepted the arguments that the architects made concerning the synthesis. Those who did were part of the wider audience of scientists; many were physicists, chemists, astronomers, and social scientists, who, in cross-linking their disciplines helped lead to the emergence of the central discipline of evolutionary biology. See also the discussion in chap. 6.

[124] See Mayr and Provine, eds., *The Evolutionary Synthesis*, for a full list of the historical actors and central texts.

[125] It was no accident that these disciplines were represented. All had been heavily institutionalized in American museums, American herbaria, and American agriculture research stations. For a history of the institutionalization of paleontology in America, see Ronald Rainger, *An Agenda for Antiquity: Henry Fairfield Osborn and Vertebrate Paleontology at the American Museum of Natural History, 1890–1935* (Tuscaloosa: University of Alabama Press, 1991).

in the 1930s. By the time of the Columbia Biological Series' publication and dissemination, the critical moment for unification—in a Machian sense—had already begun to take place. The Columbia texts, in responding to and citing each other, served to bind the heterogeneous practices of evolution and at the same time become part of the process of unification that had already started to take place in the mid-1930s. What came to be an evolutionary network cross-linking neighboring biological sciences had coalesced. Taken from the local audience of evolutionists, now linked in a complex reticulum, to the wider audience of scientists, Dobzhansky's evolutionary genetics would be used to bind further, situate, and sustain biology within the positivistic ordering of knowledge. The network that thus coalesced around evolutionary genetics, which had been grounded in genetics and ultimately in physics and chemistry and a mathematical logic, was to become even further extended, linking an even greater heterogeneous assemblage of practices. The publication of Julian Huxley's *Evolution: The Modern Synthesis*—which would take evolution to an even wider audience—signaled the unification of biology in a manner that also justified the unification of science.

Evolution: The Modern Synthesis

Dobzhansky and his *Genetics and the Origin of Species* offered a mechanistic and materialistic framework stretching from the gene to the human,[126] lending some measure of independence to biology and binding together representatives of biological disciplines through representative texts. Yet the individual who did the most to promote the newly emerging sense of unity in the biological sciences and to extend this unity to the wider global community, at the same time that he offered a framework to preserve the independence of biology, was Julian Huxley.[127] Huxley's role in the "evolutionary synthesis" has been misunderstood by historians of science. A voracious reader, an international traveler, and an indefatigable promoter of science, especially biological science, he was acutely aware of the criticisms made of evolutionary and biological practice. He was especially sensitive to these criticisms, since much of his life was devoted to leading a crusade, very much in the tradition of his

---

[126] This was not, however, a completely mechanistic and materialistic framework. This became apparent during the antireductionist debates in the early 1960s.

[127] Huxley was also to promote unity in another, larger sense: in 1942 he wrote a manuscript entitled "Unity in the U.S.A."; Box 65, Papers of Julian Sorell Huxley, The Fondren Library, Rice University, Houston, Texas. Political unity and global unity were central concerns for Huxley (see the discussion below).

grandfather Thomas Henry Huxley, to ground a humanistic philosophy in evolution. If one were to commit to a materialistic and mechanistic philosophical framework, as both Huxleys had, then constructing an ethical system as well as a meaningful existence for "Man" would have to come from some variant of *progressive* evolution. Julian Huxley's vision of such an "evolutionary humanism" was the central feature of his world-view and of his scientific endeavors.[128]

Belief in progress—an Enlightenment ideal—had been hard to sustain in the modern world, however, given the bloody aftermath of the First World War, the widespread sense of cultural degeneration, and the growing belief in the decline of the West. A global sense of fragmentation had ensued. With the rise of collective movements like communism, fascism, and Nazism, and with the onset of the Great Depression, the drive to ground an ethical system within a progressive, optimistic, and coherent worldview that gave a measure of autonomy to the individual intensified in the 1930s. For Huxley, a grounding in evolution and the construction of an evolutionary humanism became an imperative for the future of "modern man." From its inception, Huxley's major contribution to the growing literature on evolution, *Evolution: The Modern Synthesis*, was also to act as remedy for the ills of the modern world; avowedly progressive, liberal, and internationalist, it was Huxley's own ideological "testament of youth."

The decline in evolutionary studies had alarmed Huxley in part because it undermined his evolutionary humanism and his progressive worldview. The purification of evolution, its experimentalization, and the establishment of biology's own fundamental principles had early been a central feature of Huxley's lifework. Not only had he been an active contributor to the literature of evolution, but he was also instrumental in

[128] See the insightful article by John C. Greene: "The Interaction of Science and World View in Sir Julian Huxley's Evolutionary Biology," *Journal of the History of Biology* 23 (1990): 39–55. See also John C. Greene, "The History of Ideas Revisited." For a historical discussion of Julian Huxley, see C. Kenneth Waters and Albert Van Helden, eds., *Julian Huxley: Biologist and Statesman of Science* (Houston: Rice University Press, 1992). See also Colin Divall, "Capitalising on 'Science': Philosophical Ambiguity in Julian Huxley's Politics, 1920–1950" (Ph.D. diss., University of Manchester, 1985); and Marc Swetlitz, "Julian Huxley, George Gaylord Simpson and the Idea of Progress in 20th-Century Evolutionary Biology" (Ph.D. diss., University of Chicago, 1991). See also Krishna Dronamraju, *If I Am to Be Remembered: The Life and Work of Julian Huxley, with Selected Correspondence* (River Edge, N.J.: World Scientific, 1993). A special volume celebrating the centenary of Julian Huxley's birth contained historical assessment of some of Huxley's work: see especially John R. Durant, "Julian Huxley and the Development of Evolutionary Studies," in Milo Keynes and G. Ainsworth Harrison, eds., *Evolutionary Studies: A Centenary Celebration of the Life of Julian Huxley, Proceedings of the Twenty-Fourth Annual Symposium of the Eugenics Society, London, 1987* (London: Macmillan, 1989), pp. 26–40.

helping found the Society for Experimental Biology in 1925,[129] as well as collaborating with H. G. Wells on an ambitious project to lay the groundwork for *The Science of Life*.[130] By the mid-1930s Huxley was even more actively reading and contributing to the literature in evolutionary studies (although his early training was in embryology) and had been in dialogue with his American colleagues Dobzhansky and Mayr.[131] In Britain, he was in close contact with other evolutionists like Haldane, C. D. Darlington, and C. H. Waddington, but was in even closer contact with Fisher and especially Ford. The other individual who inspired Huxley was Thomas Hunt Morgan, to whom he dedicated his book.[132] In 1936 Huxley extended his interest in evolution by supporting systematics when he helped organize the Association for the Study of Systematics in Relation to General Biology. This increased his interactions with other British systematists who were trying to bring about consensus in the systematics community between the newer sciences of genetics and ecology and the older aims of taxonomy within an evolutionary framework. For Huxley and other systematists, such a new synthesis, which made the "new" systematics a "branch of general biology," made it possible to detect "evolution at work."[133] The result of these efforts to reform systematics and evolution culminated in Huxley's vision of the "new systematics," a new science, which was becoming one of the "focal points" of the biological sciences. Though he positioned himself as an "outsider" to systematics, he accepted a request from the Committee on Publications of the Association for the Study of Systematics in relation to General Biology to edit a volume entitled *The New Systematics*, which would aid

[129] For a history of the Society for Experimental Biology, see M. A. Sleigh and J. F. Sutcliffe, "The Origins and History of the Society for Experimental Biology (Comprising The Origins of Society by Lancelot Hogben, F. R. S. Aspects of the History of the Society [1923–1966])," catalogued with the Huxley Papers.

[130] See H. G. Wells, J. S. Huxley, and G. P. Wells, *The Science of Life*, 2 vols. (Garden City, N.Y.: Doubleday, 1931). H. G. Wells had also striven to create a unified worldview similar to Huxley's. In a letter to James Joyce dated 23 November 1928, he expressed his critical view of Joyce's lack of coherence in his work: "Your training has been Catholic, Irish, insurrectionary; mine, such as it was, was scientific, constructive and, I suppose, English. The frame of my mind is a world wherein a big unifying and concentrating process is possible (increase of power and range by economy and concentration of effort), a *progress* not inevitable but interesting and possible." He later added to Joyce: "Your work is an extraordinary experiment and I would go out of my way to save it from destruction or restrictive interruption. It has its believers and its following. Let them rejoice in it. To me it is a dead end." As cited in Richard Ellmann, *James Joyce: New and Revised Edition* (Oxford: Oxford University Press, 1983), p. 607.

[131] Huxley made repeated visits to Mayr and discussed chapters of *Evolution: The Modern Synthesis* prior to its publication.

[132] The dedication reads: "Dedicated to T. H. Morgan: many-sided leader in biology's advance."

[133] Julian Huxley, ed., *The New Systematics* (Oxford: Oxford University Press, 1940), p. 2.

the birth of the new science.[134] Huxley described the history of this science in his introductory "Towards the New Systematics":

> Even a quarter of a century ago it was possible to think of systematics as a specialized, rather narrow branch of biology, on the whole empirical and lacking in unifying principles, indispensable as a basis for all biological workers, but without much general interest or application to other branches of their science. Today, on the other hand, systematics has become one of the focal points of biology. Here we can check our theories concerning selection and gene-spread against concrete instances, find material for innumerable experiments, build up new inductions: the world is our laboratory, evolution itself our guinea-pig.[135]

Huxley's involvement with biologists and especially evolutionists was extensive. If any one person in the 1930s could summarize the modern evolutionary "state of the art" in palatable form for a wide audience of readers, that person was Julian Huxley.[136]

The opportunity to publish a synthetic work that would draw together the burgeoning literature in evolutionary studies within his evolutionary humanism came with the request to give the presidential address to the zoology section of the British Association for the Advancement of Science in 1936. In the essay written for this occasion Huxley expressed his wish for a unified biology and his observation that a move toward unification was taking place: "Biology at the present time is embarked upon a phase of synthesis after a period in which new disciplines were taken up in turn and worked out in comparative isolation. Nowhere is this movement towards unification more likely to be fruitful than in the many-sided topic of evolution; and already we are seeing its firstfruits in the reanimation of Darwinism which is such a striking feature of post-war biology."[137] This essay, entitled "Natural Selection and Evolutionary Progress," formed the basis for Huxley's 1942 book. From chapter 1, "The Theory of Natural Selection," which raised the status of natural selection to a theory (for Fisher, it will be recalled, natural selection was only a *genetical* theory), to the final chapter, "Evolutionary Progress,"

---

[134] Huxley positioned himself as an "outsider" in his "Foreword" to the volume. He justified his involvement in the project not only because he was "deeply interested" but also because he believed that sometimes an outsider may "be able to hold the balance between divergent views more easily than one who is himself in the thick of the fray."

[135] Ibid., pp. 1–2.

[136] J. B. S. Haldane was another popular British science writer.

[137] J. S. Huxley, "Natural Selection and Evolutionary Progress," presidential address at annual meeting, in *Report of the British Association for the Advancement of Science* 106 (1936): 81–100.

natural selection was used to ground a progressive vision of the world within Huxley's evolutionary humanism.

The evolutionary framework that Huxley adopted was adaptationist and strongly selectionist and bore the imprint of Fisherian evolution, even though he acknowledged the importance of random genetic drift and what he called the "Sewall Wright Phenomenon," and also acknowledged nonadaptive evolution. Huxley's treatment of the current state of evolutionary studies represented the diverse sets of data from international workers, paid attention to evolutionary phenomena in nonanimal systems, and took into account evolution at different levels (genic, chromosomal, individual, etc.). Hence the book appeared to consider all evolutionary points of view—yet the very structure of the book revealed Huxley's intent to ground a progressive evolution in natural selection.

Writing what would become the standard for disciplinary histories of modern evolution (and formulating the central problem for subsequent historians of biology), Huxley introduced his book by describing the woeful state of evolutionary studies that had accompanied the rise of Mendelism and the eclipse of Darwin. Chapter 1, which ostensibly was to discuss "the theory of natural selection," became a historical account of the philosophical and methodological struggle to lend scientific legitimacy to evolution. Singling out evolution as the "most central and most important of the problems of biology," Huxley urged his readers to "attack" this problem with "facts and methods" from every branch of the sciences. Biology, he recognized, was embarking on a "phase of synthesis" to bring biological and other scientific disciplines together. Aware of what he referred to (in a rather breezy manner) as the recent "movement towards unification,"[138] he further urged his readers to turn all of the scientific disciplines on this most central and most problematic

---

[138] I would not here define Huxley as a "positivist," nor claim that he had been "influenced" by the logical positivists; rather, I am expressing parallel concerns with the methodology of science in a positivistic manner. According to research, Huxley appears not to have interacted formally with the Unity of Science Movement. Very possibly there were intermediary contacts, however, especially since Huxley was gregarious and traveled widely in intellectual circles (see, for instance, the various groupings of theoretical biologists in Britain that Pnina-Abir Am has studied). He had been in very close touch with Bertrand Russell from at least 1919, and he drafted his lecture notes on "The Principles of Biology" at the same time that he began to correspond with Russell. Parallel developments in Huxley's and Russell's views of scientific principles deserve close examination, especially in light of Huxley's early interest in the antipositivism of Henri Bergson. Huxley had in his possession an autographed copy of Charles Morris's 1956 book *Varieties of Human Value* (Chicago: University of Chicago Press, 1956). The copy was signed "With warm regards, Charles Morris." Indirectly, we may at least infer that Huxley was on somewhat cordial terms with Morris. Huxley's copy of *Varieties of Human Value* is catalogued with the Huxley Papers.

"many-sided topic of evolution." Discussing the criticisms made of Darwinian methodology, and summarizing Darwin's argument for natural selection as based on "three observable facts of nature" and "two deductions from them," Huxley made it clear to the reader that natural selection—based on a deductive logical step, for Darwin—was now a "fact of nature capable of verification by observation and experiment." For Huxley, natural selection itself—through the work of evolutionists in the early 1930s—had become one of the fundamental principles of biology. With this fundamental principle, now capable of verification through observation and experiment, evolution and Darwinism were "reborn" like a "mutated phoenix risen from the ashes of the pyre." Natural selection would become an even firmer foundation for grounding Huxley's progressive evolution. The central problem of the biological sciences—evolution—was subject to the methodology of observation and experiment, and had become a rigorous science in its own right.[139]

Most important was the unified picture of biology that Huxley presented to his readers. For Huxley, the unification of biology (that great struggle for Woodger's generation) had begun to take place in the twenty-year interval preceding his own account; this "period of synthesis" had led to a science that could, in his mind, rival the unity and legitimacy of physics.[140] This "more unified science" had, in turn, made possible the rebirth of Darwinism. In a revealing passage, Huxley summarized the tale of the unification of biology and this rebirth of Darwinism:

> Biology in the last twenty years, after a period in which new disciplines were taken up in turn and worked out in comparative isolation, has become a more unified science. It has embarked upon a period of synthesis, until to-day it no longer presents the spectacle of a number of semi-independent and largely contradictory sub-sciences, but is coming to rival the unity of older sciences like physics, in which advance in any one branch leads almost at once to advance in all other fields, and theory and experiment march hand-in-hand. As one chief result, there has been a rebirth of Darwinism.[141]

[139] Huxley, *Evolution*, pp. 13–28.

[140] To biologists, physics and chemistry appeared to be unified sciences. How physicists and chemists perceived their disciplines is a separate issue.

[141] Huxley, *Evolution*, p. 26. A portion of this quotation echoes a similar passage in Neurath's work: "The new *Encyclopedia* so aims to integrate the scientific disciplines, so as to unify them, so as to dovetail them together, that advances in one will bring about advances in the others" (Otto Neurath, "Unified Science as Encyclopedic Integration," in Otto Neurath, Rudolph Carnap, and Charles Morris, eds., *International Encyclopedia of Unified Science*, vol. 1 [Chicago: University of Chicago Press, 1938], p. 24).

While Huxley's words indicated that the unification of the biological sciences was taking place, giving biology a unity and methodological legitimacy through theory and experiment that could rival that of physics, Huxley was also to construct a framework that would keep biology an independent yet unified biological science. It was through his version of progressive evolution that the delicate balance between unity and autonomy, mechanism and vitalism—the same fine line that Woodger, Haldane, and other biologists had walked earlier in the twentieth century—would be achieved. This was to come from his reborn Darwinism, progressive evolution by means of natural selection.

Given that Huxley endorsed a view of natural selection as a mechanistic principle, however, belief in evolutionary progress was exceedingly hard to sustain. If selection were strictly mechanical and nonteleological, then one could not ascribe purposiveness or directionality to evolution. No evolutionary progress would occur if there were no goal or end point for evolution. Articulating as nonteleological a version of natural selection that could still somehow give direction and make possible progressive evolution, and at the same time adhering to selection as a mechanistic—and therefore legitimate—scientific principle was the challenge that Huxley faced in the final chapter of *Evolution: The Modern Synthesis*.[142]

The manner in which Huxley articulated a progressive evolution grounded in natural selection was complex. By closely linking evolutionary progress with technological progress—humans were, after all, "unique" in their capability to modify their environment for their own purposes—he claimed that the human would in the same manner be able to control its own development through the conscious, willful use of its intelligence.[143] This same intelligence would also be able to generate human values, giving rise to "morality, pure intellect, aesthetics, and creative activity"; "man" was therefore unlike any other animal forms. In this way human improvement, and autonomy for the individual, were all combined in Huxley's vision of progressive evolution; and humans through this progressive evolution had been selected as the "unique" and "highest" of all organisms on earth.[144] Evolution, thus, was "as much a

[142] See Provine, "Progress in Evolution and Meaning in Life," for a discussion of evolution, progress, and Julian Huxley.

[143] For Huxley, "the most startling potentiality revealed by evolution" was mind, which was not contained in matter but in some form of "world-stuff." For a fascinating discussion of Huxley's theory of mind, see his contribution to the "Jubilee" celebration of genetics in 1950, "Genetics, Evolution and Human Destiny," p. 604.

[144] This position is also summarized in a manuscript of 1949 entitled "Evolutionary Humanism," Box 67.8, Huxley Papers. This was one of Huxley's recurring themes.

product of blind forces as is the falling of a stone to earth or the ebb and flow of the tides," but purpose itself would only come from human will: "If we wish to work towards a purpose for the future of man, we must formulate that purpose ourselves. Purposes in life are made, not found."[145] Evolutionary progress, in this form at least, was as mechanistic and nonpurposive as possible, yet made room enough for free will.

The framework Huxley provided was to do one more thing: it would help stabilize an ideology. The last two pages of *Evolution: The Modern Synthesis* reveal the inextricably culturally embedded features of that framework. Echoing George Orwell and Eugène Ionesco, Huxley revealed his fears of the great collectives of the 1930s that threatened to lead to the "subordination of the individual" and of leading a life whose purpose would be fulfilled in "a supernatural world."[146] The struggle between these two opposing extremes was the struggle Huxley saw facing the modern world. To provide solutions to these global problems, at the same time that he resolved the "central" problem of evolution, was his hope in upholding a progressive view of evolution. With selection acting on the individual level, the individual could be "unique" at the same time that it existed in a social group. Neither totally mechanistic/materialistic (hence avoiding the politically extreme left wing of atheists and communists), nor too vitalistic/spiritual/mystical (hence avoiding the extreme right wing of fascists, Nazis, and religious fundamentalists), Huxley's evolutionary framework balanced just enough mechanism and materialism with purpose and progress to sustain and justify a moderate liberal ideology. The threat to evolution that Huxley attributed (at the end of his first chapter) to the extremes of Henri Bergson, the ultravitalistic metaphysician, and William Bateson, the ultramechanistic materialist, was thus to be neutralized by a mechanistic yet purposive view of evolution.[147] With the end of World War II imminent, Huxley and his purpo-

---

[145] Huxley, *Evolution*, p. 576.

[146] Ibid., p. 578. Concern with the metaphysical features of life had been especially apparent in Huxley's early work; see n. 44 for notice of Bergson's influence on Huxley.

[147] See Grogin, *The Bergsonian Controversy in France*, for an account of the "revolt from mechanism." Bateson's well-known book of 1894 with the title *Materials for the Study of Variation* is an indicator of the frustration that Bateson had encountered when he could not reconcile the causo-mechanical agent of selection with any material basis for variation. His book was meant to instruct workers to search for this basis. His excitement with the "rediscovery of Mendel" was due to his seeing the material basis for variation aligned with a mechanism for evolutionary change that behaved in "law-like" fashion. Bateson, like his contemporary Hugo de Vries, had been working with plant material and focusing on the species—not clearly defined—as the unit of evolution. Hence both men were to uphold strongly saltationist points of view, given the model "planty" organisms they had adopted. Such saltationist points of view at the turn of the century were enormously popular with geneticists, many of whom had been converted from practical plant breeders housed in

sive, selectionist framework would in turn sustain an increasingly moderate, popular liberal ideology with a view of an independent biological science.

Loaded with a political and ethical perspective couched in terms of the recent developments of science, Huxley's book was not favorably received by the most local audience of evolutionists.[148] Unlike Dobzhansky's *Genetics and the Origin of Species*, Huxley's book lacked technical detail and originality, and appeared somewhat disorganized. Unlike the Columbia texts, it did not serve to bind together the heterogeneous practices of the biological sciences; rather, with its grounding in selection and its argument for a unified science of biology, it helped take this "modern" evolution to the wider audience of scientists—physical, biological, and social—in the 1940s and 1950s. For this wider audience, Huxley was to use the unified biological sciences, now "modernized" and rivaling both physics and chemistry, to help extend and legitimate both evolution and biology. Most important, with its emphasis on evolutionary progress, Huxley's book offered an inquiry—similar to his grandfather's[149]—into an ethical system, an ethos, grounded in evolution, now a legitimate science, with its fundamental principle of natural selection, verifiable and testable through observation and experiment. Evolution was portrayed as a science as mechanistic and materialistic as possible, yet preserving enough purposiveness to lend independence to the human and to biology, and at the same time to generate a measure of faith and goodness to a world grown weary of global disturbances.

As the horrors of the Holocaust, the Cold War, and the nuclear nightmare were made more apparent, the belief in selection and the adaptability of life as Huxley and others were articulating it—offering a sense of progress, a liberal ideology, and an optimistic and coherent worldview with humans as the agents of their own evolution—intensified yet further. For evolutionists living within such a world, only

horticultural and agricultural institutions. Bateson's initial address on genetics, as well as his announcement of the rediscovery of Mendel, was to an audience of the Royal Horticultural Society. Bateson—especially as discipline builder of genetics—deserves close reconsideration by historians of science.

[148] See, for instance, Dobzhansky's somewhat unflattering comments in his "Foreword" to the 1959 English translation of Bernhard Rensch's *Evolution above the Species Level*. Dobzhansky stated that Huxley's "attempt" to synthesize evolution "fell rather short of synthesis."

[149] While the two Huxleys shared certain fundamental assumptions about evolution and ethics, their formulated ethical systems were different. It will also be recalled that T. H. Huxley was not a strong supporter of natural selection. See John C. Greene, "From Huxley to Huxley: Transformations in the Darwinian Credo," chap. in *Science, Ideology, and World View* (Berkeley: University of California Press, 1981, pp. 158–93).

through some form of evolutionary humanism would human "improvement" be thought possible.[150] Such a view would also help account for, justify, and accelerate the unsurpassed success and inexorable progress of atomic age, and then space age, technology.[151] By the early 1960s—the same historical moment characterized as the most prosperous and optimistic in recent American history—the belief in selection culminated in the most extreme panselectionist doctrines. Fueled, in part, by the work of Oxford ecological geneticists A. J. Cain and P. M. Sheppard and their experiments on banding patterns in *Cepea nemoralis*,[152] and by H. B. D. Kettlewell's confirmation of industrial melanism in *Biston betularia*,[153] a "hardening" of evolution around a strongly selectionist framework took place.[154] At the same time, such a strongly selectionist framework that allowed for adaptation to a rapidly shifting technological environment would give the human individual the will and the hope to "survive."[155]

Nor was Huxley the only evolutionist to uphold a view of evolutionary progress in the 1940s. Dobzhansky, Mayr, Simpson, and Stebbins all came to subscribe to versions of evolutionary progress at the same time

[150] The word "eugenics" began to be purged from biologists' vocabularies well after the horrors of the Holocaust and Nazi medicine were made apparent. Huxley continued to use this term until at least the 1950s. Others, like Dobzhansky, continued to use the word well into the 1960s.

[151] Especially in the United States. The sense of easy progress and optimism that characterized postwar American culture was not mirrored by the war-torn Continent. This accounts for the view that the evolutionary synthesis was primarily an American (to some extent, an Anglo-American) phenomenon. The shift in evolutionary studies from Europe to the United States after the war was also reinforced by the founding of the Society for the Study of Evolution in the United States and the deliberately international journal of evolution, *Evolution*. The need to prove that American workers were as good as the Europeans in evolutionary science was one motive in founding an international journal with a base in the United States. See Smocovitis, "Organizing Evolution." Huxley had organized a British society that stressed systematics, but this organization was not as broad in its scope as the SSE. For discussion of evolutionary workers in Germany, see the work of Jonathan Harwood, especially *Styles of Scientific Thought: The German Genetics Community, 1900–1933* (Chicago: University of Chicago Press, 1993); "Geneticists and the Evolutionary Synthesis in Interwar Germany," *Annals of Science* 42 (1985): 279–301; and "Metaphysical Foundations of the Evolutionary Synthesis: A Historiographical Note," *Journal of the History of Biology* 27 (1994): 1–20.

[152] A. J. Cain and P. M. Sheppard, "Natural Selection in *Cepea*," *Genetics* 39 (1954): 89–116.

[153] H. B. D. Kettlewell, "Selection Experiments on Industrial Melanism in the *Lepidoptera*," *Heredity* 9 (1955): 323–42.

[154] See Stephen Jay Gould, "The Hardening of the Modern Synthesis," in Marjorie Grene, ed., *Dimensions of Darwinism* (Cambridge: Cambridge University Press, 1983), pp. 71–93.

[155] The themes of survival and human improvement, the uniqueness of the human mind, and the path of biological progress played themselves out in much of Huxley's work. See his *Evolution in Action* (New York: Harper and Brothers, 1953).

that they made natural selection a mechanism.[156] Each of these evolutionists in some measure addressed the "future of Mankind," either in con-
cluding chapters of their early books or in more popular books written at
later stages of their careers.[157] Echoing Huxley's evolutionary humanism,
Simpson wrote: "It is another unique quality of man that he, for the first
time in the history of life, has increasing power to choose his course and
to influence his own future evolution. It would be rash, indeed, to attempt to predict his choice. The possibility of choice can be shown to
exist. This makes rational the hope that choice may sometime lead to
what is good and right for man. Responsibility for defining and for seeking that end belongs to all of us."[158]

The belief in the continuum between the gene and the human brought
these humanistic concerns within the materialistic and mechanistic frameworks of genetics and the physical sciences.[159] Evolutionary progress
through the mechanism of evolution, though it appeared to be a contradiction, struck just the right balance among purpose, progress, and
mechanistic materialism—it was deterministic enough to be predictable,
yet not enough to remove free will—for a wide audience of evolutionists,
who took their views through textbooks, semipopular works, and essays
to the wider audience of scientists.[160]

In the late 1950s and early 1960s popular culture—permeated with
evolutionary science—resonated with evolutionary themes. As science
fiction, an emerging literary genre, simultaneously burgeoned at this

[156] See Nitecki, *Evolutionary Progress*, for an indication of how contentious the subject of
evolution and progress has been. See also Marc Swetlitz, "Julian Huxley, George Gaylord
Simpson and the Idea of Progress in 20th-Century Evolutionary Biology" (Ph.D. diss.,
University of Chicago, 1991).

[157] See, for instance, Theodosius Dobzhansky, *Mankind Evolving: The Evolution of the
Human Species* (New Haven: Yale University Press, 1962); idem, *The Biology of Ultimate
Concern* (New York: World, 1967); and Dobzhansky's "Introduction" to the "Life Nature
Library series" volume: Ruth Moore, ed., *Evolution* (New York: Time-Life Books, 1962);
see also George Gaylord Simpson, *The Meaning of Evolution* (New Haven: Yale University
Press, 1949); idem, *This View of Life: The World of an Evolutionist* (New York: Harcourt
Brace Jovanovich, 1964); idem, *Biology and Man* (New York: Harcourt, Brace and World,
1969).

[158] Simpson, *The Meaning of Evolution*, p. 348.

[159] Punctuated equilibrium and the critique of the adaptationist program, launched by
Stephen Jay Gould and Richard C. Lewontin, was to construct an argument that would
lead to a sundering of this continuum; see S. J. Gould and R. C. Lewontin, "The Spandrels
of San Marco and the Panglossian Paradigm: A Critique of the Adaptationist Programme,"
Proceedings of the Royal Society of London B 205 (1979): 581–98. For a response to the
Gould and Lewontin argument, see Ernst Mayr, "How to Carry Out the Adaptationist
Program?" *American Naturalist* 121 (1983): 324–34.

[160] Another example of this balance among evolutionary progress, purpose, and mechanistic materialism in semipopular form is the published book of Dobzhansky's 1961 Silli-
man Lectures, *Mankind Evolving*.

time, the themes of evolutionary progress and technological progress leading to futuristic hyperintelligent forms of life became a commonplace for popular audiences.[161] Technological progress and evolutionary progress, inextricably linked, were to hold the key to the future of the most intelligent and unique species on earth: "modern man."

For practitioners of biology who sought legitimacy and wished to avoid engulfment from the physical sciences, evolution made an independent science of life defensible. Evolution as a mechanistic and materialistic science preserved enough teleological components through evolutionary progress to give the science of life—and the human—independence from the physical sciences. Natural selection, the primary mechanism of evolution, had just the right measure of the mechanical and the teleological: biology was preserved as an independent science with enough agency for the human, but a science that could still rival physics and chemistry through observation and experiment.

The middle ground that Woodger and other biologists had sought in the 1920s was thus found, and the principle of great "unifying power," the "generalization" that would "knit" the "several branches" into a whole biology, now existed—problematic because it contained some metaphysical elements, but just enough metaphysical properties to lend independence to biology. Experimental and quantitative in its own right, the emerging science of evolutionary biology would serve to unify and

---

[161] One of the most widely read science fiction novels written at this time represents these same themes: Arthur C. Clarke, *Childhood's End* (New York: Ballantine, 1953). Clarke drew heavily on the work of British science fiction writer Olaf Stapledon, whose lifework embodied the search for a meaningful modernist worldview within evolutionary themes. Stapledon, in turn, had drawn heavily on H. G. Wells. Wells had collaborated with Julian Huxley on the *Science of Life* and had been part of a social circle that included another very influential British science fiction writer: Aldous Huxley, the brother of Julian. For an analysis of some of these recurring themes and the social circle of science fiction writers, see Brian W. Aldiss, *Billion Year Spree: The True History of Science Fiction* (Garden City, N.J.: Doubleday, 1973). According to Aldiss and historians of science fiction, technological progress, evolutionary progress, human "improvement," and the control of human destiny were linked even in the earliest genres of science fiction, like Mary Shelley's *Frankenstein*. In the 1960s the science fiction genre made its way to television audiences. "The Outer Limits," one of the more popular of these science fiction series, frequently drew on the themes of evolutionary progress and technological progress. An episode entitled "The Sixth Finger" represented these themes. Through the creation of a special machine that sped up "mankind's evolution," a new and advanced human with increased cranial capacity was created. Along with this increase in intelligence, the advanced human bore a sixth finger, a symbolic representation of increased dexterity. This human clearly became a hyperhuman in its increased intelligence, but at the expense of its own humanity. Hence, the theme of the episode suggested that caution should be exercised lest pronounced evolutionary intervention lead to loss of "humanity." "The Sixth Finger," *The Outer Limits*, prod. Joseph Stefano, dir. James Goldstone, writ. Ellis St. Joseph, Daystar-Villa Di Stefano UA, 1963. MGM/UA Home Video, 1987. Videocassette.

bind the heterogeneous practices of the biological sciences at the same time that it would "lift" biology from reduction to the physical sciences. Evolution would eventually become the "central science" of biology.

But for the wider audience, evolution by means of natural selection, as Huxley had been promoting it, was to become a fact.[162] The publication of Huxley's book—along with his numerous essays, speeches, reviews, and encyclopedia entries—was to connect and cross-link the network that had coalesced around Dobzhansky, with Fisher and Ford, and in turn with the wider audience of scientists. As the war ended in 1945 and communication networks, extensively developed during wartime, were redirected to peacetime operations, and as the postwar optimism and accumulation of economic and material resources (in the United States) were redirected toward reconstructing and stabilizing the global order, the cross-linking of the networks of scientific practice accelerated further. A compound-complex reticulum connecting the disciplines of the sciences had emerged. The heterogeneity of not only biology, but also science, would now be bound by the central science of biology, reducible to the level of the gene and obeying the axiomatic Hardy–Weinberg Equilibrium Principle.[163] Biology, occupying the midpoint in this continuum from the physical world of the gene to the social and cultural, was the domain that would provide the most information and hold the key to the future of "modern man"—the object of study of both the biological and social sciences.[164]

The Enlightenment ideals of the proper systematic study of "Man," culminating in evolutionary humanism, liberalism, progress, and the unity of science and of all knowledge, would hold sway by the early 1950s. Knowledge would be unified by reduction to the physical sciences, whereas the diversity and variety of knowledge would emerge from the social sciences above. Evolution, partly reducible *to* the physical

---

[162] For Darwin's readers only the facticity of evolution had been established. The causo-mechanical agent had not been demonstrated.

[163] For this reason the sense arose that mathematical population genetics forms the "core" of evolutionary theory. See, for instance, how one of the earliest histories of genetics supports this. L. C. Dunn wrote that

Population genetics has come to occupy a rather special place in biology. It represents the interest in the processes of evolution and in the improvement of domesticated plants and animals which strongly motivated the early students of genetics. It made possible the great renaissance of evolutionary biology which began about 1930. Population genetics then tended to reunite fields of biology such as genetics, ecology, paleontology, and systematics, which had tended to take separate paths. It has thus been referred to, and with cause, as the core subject of general biology. (L. C. Dunn, *A Short History of Genetics* [New York: McGraw-Hill, 1965], p. 192)

[164] Once again accounting for the nature–nurture debates that flared after this time.

world, but also emergent *from* the physical world, would lead ultimately to the progressive divergence of knowledge. The "growth" of scientific knowledge was thus to take the trajectory of a progressively diverging path. The ever-branching, ever-ramifying "tree of life" began to map a one-to-one correspondence with the ever-branching, ever-ramifying "tree of knowledge." Bearing special signification for religious systems of thought,[165] this metaphor herein represented an end to conventional Judeo-Christian thought: a secular, yet meaningful evolutionary humanism had thus emerged.

The "Unity of Knowledge" was the name of a conference with the central theme of "Man's Right to Knowledge and the Free Use Thereof," held in honor of the bicentennial of Columbia University in 1954.[166] The list of contributors to, and participants in, this conference included luminaries from all the existing disciplines of knowledge: Pierre Teilhard de Chardin, Theodosius Dobzhansky, Julian Huxley, Talcott Parsons, B. F. Skinner, Harold Urey, Niels Bohr, John Von Neumann, Willard Van

[165] For biologists the tree of life is associated with Charles Darwin's evolutionary theory, but the metaphor of the tree of life runs through many cultures and is frequently associated with religious and mystical themes. This is discussed in Roger Cook, *The Tree of Life: Image for the Cosmos* (New York: Thames and Hudson, 1974; repr. 1992); Roger Malbert (exhibition organizer), *The Tree of Life: New Images of an Ancient Symbol* (London: South Bank Board, 1989); see also the entry "Tree of Life," in *Man, Myth and Magic* (New York: Marshall Cavendish, 1995). Here I am using the metaphor to represent the end of conventional Western Judeo-Christian religion. According to Alfred Ernest Crawley, the alignment of the tree of life with the tree of knowledge represented the crucifixion according to medieval mysticism. Crawley maintains that the tree of life represents the body of Christ, and the tree of knowledge represents the crucifix. See Alfred Ernest Crawley, *The Tree of Life: A Study of Religion* (London: Hutchinson, 1905). The religious and mystical tone of the metaphor of the tree of life runs through many world religions but is especially evident in Judaism and Buddhism. In Judaism it is associated with the Bible or the Wisdom Literature because it deals directly with life: see Roland E. Murphy, *The Tree of Life: An Exploration of Biblical Wisdom Literature* (Garden City, N.Y.: Doubleday, 1990). See also the teachings of Rabbi Louis Hammer, *The Tree of Life: A Guide for Youth* (Brooklyn: The Reader's Guide and Library Supply, 1937). For Hammer the tree of life represents the Bible and the tree of knowledge represents science and technology. Modern man's acceptance of science and technology means a rejection of the tree of life (the Bible). The metaphor of the tree of life also runs deeply in Eastern philosophy. For its use in Buddhism, see *Tree of Life: Buddhism and Protection of Nature, with a Declaration on Environmental Ethics from His Holiness The Dalai Lama* (Buddhist Perception of Nature, 1987). More recently the metaphor of the tree of knowledge has made its way into biological bases of cognition in quasi-holistic form. See Humberto R. Maturana and Francisco J. Varela, *The Tree of Knowledge: The Biological Roots of Human Understanding*, rev. ed. (Boston: Shambala, 1992).

[166] The conference took place at Arden House, Harriman, New York, on 27–30 October 1954. Members of the planning committee included Albert Hofstadter, Paul Lang, Ernest Nagel, Marjorie Hope Nicolson, I. I. Rabi, and Lionel Trilling. The panels for the conference were organized and managed by Horace Friess, Ernest Nagel, and Jacques Barzun. Also helping with the general sessions were John A. Krout, Henry P. Van Duesen, Edgar Grim Miller, and Philip C. Jessup. For details of the conference and the edited proceedings, see Lewis Leary, ed., *The Unity of Knowledge* (Garden City, N.Y.: Doubleday, 1955).

Orman Quine, Ernest Nagel, and Philipp Frank.[167] The compound-complex reticulum, a polysemous web, linked the metaphysician with the mathematician—the physicist, the chemist, the biologist, the psychologist, and the sociologist with the philosophers and logicians of the Vienna Circle.[168] All were enveloped by the same Enlightenment ideals writ large: progress, unity, diversity, and a liberal, evolutionary humanism.[169] The title of Huxley's contributed paper, which articulated an evolutionary humanism through evolution by means of natural selection, reveals his role in the modern synthesis: the determination of "Man's Place and Role in Nature."[170]

In 1946 Huxley was to take his modern synthesis of evolution to a still wider audience: the "United Nations" of earth. Becoming director-general of UNESCO, an emerging global political force, Huxley believed his drive to unify biology within an evolutionary worldview would aid the process of unifying a fragmented world in search of a common ground for political unity. By the 1950s the "modern" synthesis of evolution had thus reached an international audience of the "modern" unified nation-states: based on liberalism, evolutionary humanism, and a belief in international progress, a new global political community had emerged.[171]

While this larger global network was striving to reach consensus on a unified theory of knowledge, the more local evolutionary network that had coalesced around evolutionary genetics had been secured by becom-

---

[167] The Leary volume includes a complete list of attendees.

[168] Philipp Frank, it will be recalled, had been a promulgator of the Machian position that the unification of the sciences had to take place through the ejection of metaphysics. See Holton, "Ernst Mach and the Fortunes of Positivism in America."

[169] The driving force behind these Enlightenment values was the Newtonian mechanistic picture, which preserved enough metaphysics for a meaningful life devoid of complete determinism. The unification of knowledge—which brought "Man" into the deterministic fold of mechanistic and materialistic genetics—was therefore the culmination of Enlightenment thought. One feature that emerged from the "Unity of Knowledge" conference was the need to exert caution with respect to political unity. This was one way to avoid the twin specters of communism and fascism. Leary's introduction also suggests that conference attendees were to modify their originally "bold concept of a unity of knowledge" to a more "manageable" and "semantically more sound" concept of the unification of knowledge. See Leary, "Preface," in *Unity of Knowledge*, p. xi.

[170] Dobzhansky was to make "Man" the "Centre of the Universe" in the final section of *Mankind Evolving*. In this final section, he made explicit his warm regard for Teilhard de Chardin; see P. Teilhard de Chardin, *Le phénomène humain* (Paris: Editions Du Seuil, 1955); English trans., *The Phenomenon of Man* (New York: Harper and Row, 1959). See also Dobzhansky's textbook of 1955 entitled *Evolution, Genetics and Man*. Here he viewed "Man" as "the Pinnacle of Evolution" (p. 373). Julian Huxley also held Teilhard de Chardin in high regard. He introduced *The Phenomenon of Man*.

[171] No surprise that Huxley learned to use effectively modern communication technology like radio and television, even becoming a radio celebrity by making regular appearances on programs like "The Brains Trust." He worked extensively for the BBC.

ing officialized and institutionalized as the central component of the biological sciences. The lifelong task of these scientists, redefined as evolutionary biologists, would be to balance the positivistic ordering of knowledge from deep within the biological sciences. Disciplining evolutionary biology—the fulcrum of the biological sciences—they were to act as unifiers, negotiators of the location of biology, preservers of the whole of the positivistic ordering of Enlightenment knowledge.

DISCIPLINING EVOLUTIONARY BIOLOGY

The local network around Dobzhansky's evolutionary genetics had coalesced rapidly—in fact, a consensus that a common ground existed arose within a few years of Dobzhansky's initial work on the genetics of natural populations. At the same moment that evolutionary studies reached their most endangered point, the new experimental and mechanistic and materialistic evolutionary practices promised to help preserve evolutionary studies. Along with the consensus that a common ground existed, there had simultaneously come the agreement that the evolutionary practices that resolved problems of speciation should be secured and sustained by being institutionalized.[172] In 1939 at a special "Symposium on Speciation" organized by Dobzhansky for the American Association for the Advancement of Science (AAAS) meetings in Columbus, Ohio, Julian Huxley met with Dobzhansky, Mayr, and Carl Epling to suggest the formation of an official Society for the Study of Speciation, which would function as an information service that could introduce interested workers to each other.[173] At the same time, Huxley was bringing together a wide group of practitioners interested in problems of speciation in Britain under the rubric of the "new systematics."[174]

Although an informal Society for the Study of Speciation that would function as an information service was officially founded in the United States in 1940 under the secretaryship of A. E. Emerson, the outbreak of the war interrupted the initial impetus and thwarted further plans. Support for an informal communication service among evolutionists continued, however. On the West Coast the "biosystematists," a San Francisco Bay area informal cooperative organization, continued to support evolutionary studies; but it was the collaboration between New York-based paleontologists, systematists, and geneticists that would exert pressure to

[172] See Smocovitis, "Organizing Evolution."
[173] Reviews and Comments, "Evolution News," *American Naturalist* 75 (1941): 86–89.
[174] See Julian Huxley, ed., *The New Systematics* (Oxford: Oxford University Press, 1940).

institutionalize evolutionary studies. In the early 1940s this latter group, at the initiative of Walter Bucher at Columbia University, made additional moves to launch a "synthetic attack" on the "common problems of evolution" by forming a cooperative and coordinated organization.[175] These plans were put into effect in 1943 when the Committee on Common Problems of Genetics, Paleontology, and Systematics was established under the auspices of the National Research Council.[176] Two meetings were held in 1943: one at the American Museum of Natural History in New York (this group was heavily represented by paleontologists and geneticists) and one at the University of California–Berkeley that drew on the "biosystematists" and botanists in the San Francisco Bay area.

Although the committee heartily supported the extension of these local meetings to the national level, more ambitious plans were thwarted by wartime conditions. During the difficult war years, communication among evolutionists took place through a series of mimeographed bulletins edited by Ernst Mayr in which "common problems" were discussed by the local network of evolutionary practitioners.[177] The problems discussed in the bulletins indicated that the interest had expanded to represent a wider set of issues that was emerging from the synthesis in points of view. Not only were the initial problems of speciation, divergence, and isolation discussed, but also problems of evolutionary rates and higher-order phenomena. The widening of the perspective was one outcome of the presence of the New York paleontologists and West Coast botanists who wished to understand evolutionary rates and variation and evolution in plants. It was through these communication bulletins— written in a series of letters or rapid exchanges among members of the local group—that a consensus emerged that there was in fact not only a common ground but a common field that should be officialized.[178] Re-

[175] See the historical "Foreword" to the edited volume of the proceedings of the Princeton conference written by Glenn L. Jepsen, in Glenn L. Jepsen, Ernst Mayr, and George Gaylord Simpson, eds., *Genetics, Paleontology, and Evolution* (1949; repr. New York: Atheneum, 1963).

[176] This was a joint or interdivisional committee organized by the division of geology and geography and the division of biology and agriculture of the National Research Council.

[177] I have consulted the set of mimeographed volumes in the holdings of the Provine evolution collection in Marathon, New York, as well as an unpublished manuscript by Ernst Mayr, "History of the Society for the Study of Evolution" (dated most likely around 1947), included with the bundle of mimeographed bulletins. See also Smocovitis, "Organizing Evolution."

[178] The dialogue format of these bulletins facilitated the construction of a disciplinary discourse.

turning from military service abroad, Simpson identified the emergence of this common field in the final mimeographed bulletin of 1944:

> This series of bulletins, compiled and edited by Dr. Mayr who continues this task, has accomplished a great deal more than the expression of a few facts and opinions, useful as these have also been. From the whole series of letters in the bulletin there has emerged concrete evidence that a field common to the disciplines of genetics, paleontology, and systematics does really exist and this field is beginning to be clearly defined. Some, at least, of the more hopeful approaches to these common problems are indicated and exemplified. The existence of geneticists, paleontologists, and systematists interested in these problems and competent to attack them has been demonstrated. Their interest has been stimulated and made more concrete and their competence in the joint field has been increased by the exchange of views with students of other specialties. Thus great progress toward the goal of the committee has been made.[179]

By this time, too, the Columbia texts' representing of the heterogeneous practices of evolution was reaching the wider audience, garnering even further support and belief in the emergence of a now common field of practice. But it was only after the war that major moves could be made to redirect available resources to peacetime operations, such as the planning of major conferences and the creation of new societies. With the end of the war, what had been the defunct Society for the Study of Speciation, with the help of the Committee on Common Problems in Genetics, Paleontology, and Systematics, was recreated under the initiative of Ernst Mayr. On Saturday, 30 March 1946, in St. Louis, Missouri, fifty-eight attendees—the "founding fathers"[180]—signed a document, entering a confederacy under the title of the Society for the Study of Evolution (SSE; see fig. 2).[181] The minutes of the first organizational meeting and subsequent accounts indicate little disagreement, given the diverse set of backgrounds of the members.[182] The most contentious issue ap-

---

[179] G. G. Simpson, "Introductory Remarks," Bulletin no. 4, 13 November 1944.

[180] Ruth Patrick was the only female signatory.

[181] Alfred Emerson presided at this meeting. The first president of the SSE was G. G. Simpson, with Ernst Mayr as secretary. See Smocovitis, "Founding the Society for the Study of Evolution."

[182] The minutes of the first meeting were recorded by R. P. Wagner, "The Society for the Study of Evolution. Organization meeting, March 30, 1946," SSE Papers, Library of the American Philosophical Society.

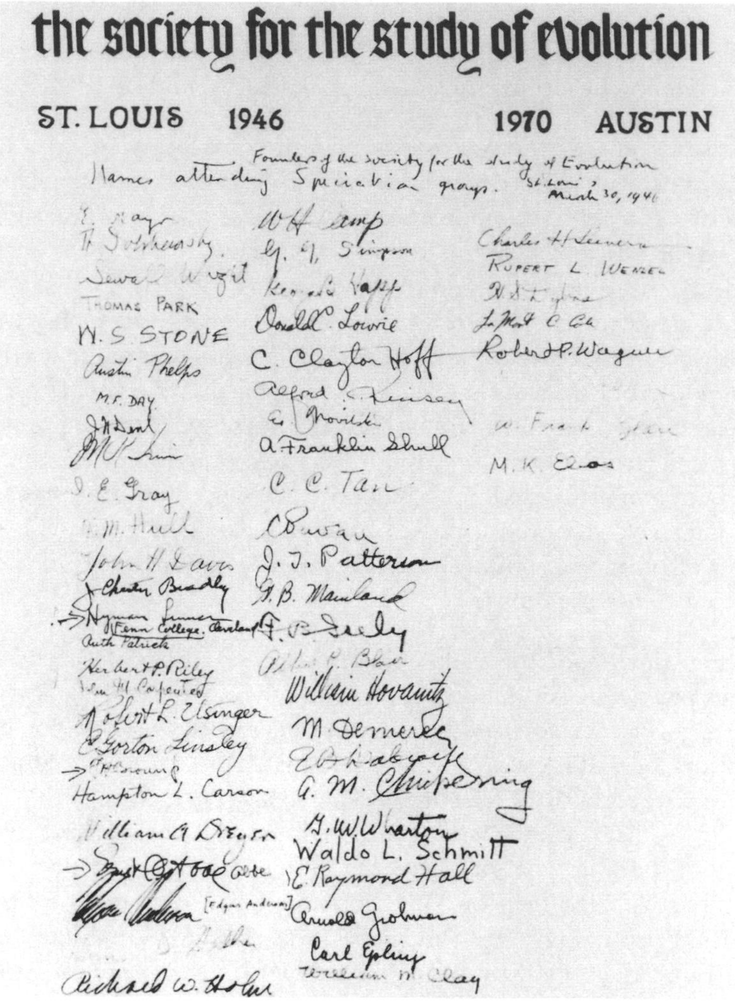

the society for the study of evolution

ST. LOUIS 1946   1970 AUSTIN

FIGURE 2: Foundation Document of the Founders of the Society for the Study of Evolution, St. Louis, 30 March, 1946.

Original List of Signatories at the 1946 Meetings in St. Louis. The small handwriting above the signatures is by Ernst Mayr. The signatures, by column from left to right are: (1) E. Mayr, Th. Dobzhansky, Sewall Wright, Thomas Park, W. S. Stone, Austin Phelps, M. F. Day, J. N. Dent, M. R. Irwin, I. E. Gray, F. M. Hull, John H. Davis, J. Chester Bradley, Hyman Lumer, Ruth Patrick, Herbert P. Riley, John M. Carpenter, Robert L. Usinger, E. Gorton Linsley, F. J. Brounp (?), Hampton L. Carson, William A. Dreyer, Ernst C. Abbe, Edgar Anderson, Harrison D. Stalker, Richard W. Holm; (2) W. H. Camp, G. G. Simpson, George B. Happ, Donald C. Lowrie, C. Clayton Hoff, Alfred Kinsey, E. Novitski, A. Franklin Shull, C. C. Tan, C. Pavan, J. T. Patterson, G. B. Mainland, F. B. Isely, Albert P. Blair, William Hovanitz, M. Demerec, E. B. Babcock, A. M. Chickering, G. W. Wharton, Waldo L. Schmitt, E. Raymond Hall, Arnold Grobman, Carl Epling, William M. Clay; (3) Charles H. Seevers, Rupert L. Wenzel, H. S. Dybas, Lamont C. Cole, Robert P. Wagner, Alfred Emerson, W. Frank Blair, M. K. Elias. The document was reproduced and distributed to members at the 1970 SSE meetings in Austin, Texas. Photograph courtesy of James Crow and Donald Waller. An additional document listing founders also includes C. W. Metz.

peared to be the relations with the *American Naturalist* (the journal of the American Society of Naturalists), which was possibly endangered by the creation of the new society that was too close in its scope.

Within the year, the first annual meeting was held in Boston (28–31 December), at which time the members of the society felt sufficiently established to begin a journal to reach an international audience of evolutionists. After deliberation, the journal, which would promote the new experimental practices rather than just descriptive or taxonomic practices, was entitled *Evolution*, with Ernst Mayr as editor.[183] Such an international journal would also help soothe war-torn, frayed nerves and increase collaboration among international evolutionists. With the support of astronomer Harlow Shapley, $5,000 in funds had come from a "Reserve Fund for Post-War Expenditures" granted through the American Philosophical Society. Ernst Mayr and G. G. Simpson had written the grant with the appeal to "unifying the fields."[184] At the Boston meeting, too, a constitution was drafted that would fix and sustain the common goals of the society.

Within a week of the drafting of the constitution at the Boston meetings, the Committee on Common Problems of Genetics, Paleontology, and Systematics held its final symposium at Princeton as part of the university's bicentennial conferences. In a spirit of celebration and rejoicing over the new consensus among formerly dissenting groups of biologists, participants agreed that "a convergence of evolutionary disciplines" had taken place.[185] Writing the summation of the edited volume of the proceedings, H. J. Muller drew the parallel between an evolutionary convergence of types and a convergence of disciplinary types like geneticists and paleontologists. The end result of this fusion was a new and higher type, through a process of synthesis: the synthetic type of evolutionist. Muller's summation, which outlined the consensus in evolutionary practice and reproblematized the now common field, indicated a shift in what was emerging as the evolutionary disciplinary problematic. The points of agreement included the primacy of natural selection as a mechanism of

[183] Ernst Mayr was instrumental in raising support for the journal, as well as being a key, if not *the* key, player in the founding of the SSE. Simpson played an important role in assisting Mayr to obtain start-up funds.

[184] As cited in Cain, "Common Problems and Cooperative Solutions."

[185] This was H. J. Muller's subtitle in the "Summation" to the Princeton volume; see H. J. Muller, "Redintegration of the Symposium on Genetics, Paleontology, and Evolution," in Jepsen, Mayr, and Simpson, eds., *Genetics, Paleontology and Evolution*, pp. 421–45. Both Ernst Mayr and G. Ledyard Stebbins sense that evolutionary biology emerged at about the same time as the Princeton meetings (letter from Ernst Mayr to author, 15 August 1989; letter from G. L. Stebbins to author, 4 May 1989).

evolutionary change, the gradual rate of change operating at the level of small, individual differences, and the continuum between microevolution and macroevolution—reconstituting what Muller viewed as the original Darwinism torn apart by "over-zealous 'mutationists'" in the geneticists' camp and paleontologists who would embrace Lamarckian inheritance and an "inner evolutionary urge." Both disciplines now had a "common ground of theory." Though his closing thoughts indicated an awareness that a mechanistic and materialistic view of evolution was nonpurposive and nonprogressive and did not bode well for the future of all species, Muller echoed Huxley's belief that humans could somehow rise above their own evolutionary destiny: "If, then, we wish evolution to proceed in ways that we consider progressive, we ourselves must become the agents that make it do so. And all our studies of evolution must finally converge in that direction."[186]

Although there appeared to be agreement on the common problems and practices leading to a common ground, there was also dissent. One individual, in particular, threatened to upset the emerging consensus from within the evolutionary ranks. Though Richard Goldschmidt was an insider in the emerging field, he was an outsider as well.[187] For Goldschmidt, who was concerned with embryology and physiology (the two areas not brought into the modern synthesis), too much emphasis had been given to the gene (especially by the Morgan group) as determiner of transmission effects, rather than looking also at the action of genes (as in their physiological effects). In his opinion, genetics without biochemistry, and understanding of such gene action within what he called a "physiological genetics,"[188] could not account for the embryological and transformational features of evolution. In focusing on such physiological effects, Goldschmidt was challenging some of the central components of Dobzhansky's synthetic framework: in espousing his own genetic "macromutation" events and regulative changes in proteins that could lead to rapid saltatory events (producing his famous "hopeful monsters"), he was

[186] Muller, "Redintegration," p. 445.
[187] Scott F. Gilbert has offered an account of Goldschmidt as "outsider": see Scott F. Gilbert, "Cellular Politics: Ernest Everett Just, Richard B. Goldschmidt, and the Attempt to Reconcile Embryology and Genetics," in Rainger, Benson, and Maienschein, eds., *The American Development of Biology*, pp. 311–46; see also Garland E. Allen, "Opposition to the Mendelian-Chromosome Theory: The Physiological and Developmental Genetics of Richard Goldschmidt," *Journal of the History of Biology* 7 (1974): 49–92. For a recent study of Goldschmidt and the synthesis, see Michael Dietrich, "Richard Goldschmidt's 'Heresies' and the Evolutionary Synthesis," *Journal of the History of Biology* 28 (1995): 431–61.
[188] The title of his 1938 book: R. Goldschmidt, *Physiological Genetics* (New York: McGraw-Hill, 1938).

counteracting the efforts to limit the rate of change to small, individual differences, worked out at the level of point mutations. Furthermore, by espousing such unpredictable mechanisms of evolutionary change leading to large-scale effects that could give rise to higher-order phenomena, Goldschmidt was also challenging the continuum between microevolution and macroevolution. In postulating different mechanisms operating at different levels of evolution he impugned not only the emerging evolutionary consensus, but the very ground on which it was being built—the classical genetics of the Morgan school. The title of Goldschmidt's 1940 book, *The Material Basis of Evolution*,[189] faintly echoing William Bateson's 1894 *Materials for the Study of Variation*, was a direct response to existing knowledge of genetics and evolution that was meant to return to originary genetical concerns—unresolvable without Goldschmidt and his physiological genetics.

For Dobzhansky, and for Mayr and others in the emerging new field, Goldschmidt posed a series of ultimate challenges that threatened to upset the delicate balance that Dobzhansky's framework had struck. Goldschmidt's focus on the biochemical and physiological components of evolution made his views too materialistic; his adoption of position effects and saltatory changes made evolution too rapid and indeterministic; and his postulation of different mechanisms at work within microevolution and macroevolution threatened to unwind and disunify the biological sciences from within. With what additionally appeared an inflammatory and what some had perceived an "obstructionist" disregard for the framework as it was being built,[190] bitter controversy erupted between Goldschmidt and Dobzhansky (whose framework was challenged) and Mayr (who would do the most to promote biology as an autonomous science); both had done a great deal to strike a balance that would bring the emerging new group to consensus. Goldschmidt's closing thoughts in his 1940 book hit an especially vulnerable spot in the delicate balance that the architects had worked so hard to strike:

> The following period of experimental biology was skeptical of, if not actually hostile to, evolution, as it could not be attacked in laboratory experimentation. Mechanism became unpopular and vitalistic and teleological trends invaded evolutionary thought in the form of creative evolution, emergent evolution, psycho-Lamarckism. The rise of ge-

[189] Richard Goldschmidt, *The Material Basis of Evolution* (1940; repr. New Haven: Yale University Press, 1982).

[190] Garland Allen discusses this "obstructionist" component of Goldschmidt's complex personality. See Allen, "Opposition to the Medelian-Chromosome Theory."

netics brought back a mechanistic attitude; evolution started to become an exact science. Just as there is no room for transcendental principles in experimental physics and chemistry, in the same way a factual
attack upon the problems of evolution can work only with simple
mechanistic principles. Genetics showed the evolutionists that evolution can be attacked scientifically only on the basis of known analyz-
able processes, which are by their very nature relatively simple. But,
just as has been the case in chemistry and physics, mechanistic analysis
of evolution will sooner or later reach a point where an interpretation
in terms of known processes will meet with difficulties. In such a situation chemistry and physics have never invoked transcendental principles on the assumption that nature is so frightfully complicated that it
cannot be understood otherwise. The actual developments have shown
that this is not the case. The modern development of the electronic
theory has shown that rather simple principles govern the most complicated phenomena of matter. Of course, there is always an unex-
plained residue on which the investigator may train his personal metaphysical predilection, but certainly no chemist would look to
metaphysics for an explanation of a difficult phenomenon, say catalysis.
In the same way the evolutionist, who meets with difficulties in mechanical interpretation at a lower level, may enjoy letting loose his
metaphysical yearnings. But as an investigator he can only work under
the assumption that a solution in terms of known laws of nature is
possible.[191]

For Goldschmidt, evolution as the architects were constructing it was
therefore too metaphysical and not materialistic enough. In pointing this
out, he threatened to destabilize the fine line that the architects were
trying to walk between the unity of the sciences and the independence of
biology: the end result for biology, as he envisioned it, would come dan-
gerously close to engulfment by the physical sciences. But the dialogue
with Goldschmidt was also to have a securing effect in that it helped
articulate and refine what had emerged as the disciplinary problematic of
evolutionary biology. In disciplining the study of evolution—through
the determination of who and what counted as "outside"—the "inside"
members of the society would also negotiate the disciplinary standards
and reconstruct the disciplinary problematic of evolutionary biology.[192]
Goldschmidt was eventually "marginalized," though he would be resur-

---

[191] Ibid., p. 397.
[192] It will be recalled that Mayr had dedicated twenty-four pages of argument in his
Systematics and the Origin of Species to counteracting Goldschmidt's views of macromuta-
tionism.

rected as a "heretic" and an antihero by the next generation of evolution-ists[193]—but well before then, evolution and biology would be fixed and sustained and secured as legitimate disciplines of knowledge.

By the late 1940s, then, the legitimation of evolutionary studies, the rebirth of Darwinism, and the emergence of an experimental biological science of evolution (evolutionary biology) had been instituted officially by the formation of a recognized scientific society, the Society for the Study of Evolution, and a textual forum for expressing concerns and unifying the emerging new field, the journal *Evolution*. Simultaneously, a realignment of biological disciplines began to take place as the group of biologists finding a "common ground" of genetics and selection theory redefined their disciplinary identities as evolutionary biologists.

Also at this time, the first umbrella-like organization for the biological sciences, the American Institute of Biological Sciences (AIBS), was formed. Emerging from the Division of Biology and Agriculture of the National Research Council, it became an independent organization con-sisting of a federation of biological societies in 1947. While both the SSE and the AIBS had benefited from postwar optimism, the boom in avail-able resources, and the call to reform American science and technology,[194] the two societies were to be closely linked in deeper ways, since evolution would eventually form the central science of the unified biological sci-ences.[195] Never reaching departmental status, nor having any one tie to

[193] It was Stephen Jay Gould and others who were to portray Goldschmidt as a "heretic" and an antihero. Gould introduced the reissue of Goldschmidt's book for Yale University Press in 1982. For Gould, who inherited problems of development as well as problems in accounting for rates of evolutionary change as made apparent in the fossil record, the posi-tion that Goldschmidt represented closely resembled his own. The sundering of the contin-uum between microevolution and macroevolution in the late 1970s and early 1980s also led to the sundering of the continuum that gave rise to sociobiology. Hence, the major amendments to evolutionary theory in the early 1980s and the emergence of paleobiology were also to sustain what (in the 1980s) was a politically moderate position. For Gould et al., the autonomy of biology and the role of evolution as the "central organizing principle" would come from an argument they inherited from Simpson—chance and contingency in the form of unique historical events. See Stephen Jay Gould, *Wonderful Life: The Burgess Shale and the Nature of History* (New York: W. W. Norton, 1989). Goldschmidt's "here-sies" are also discussed in Michael Dietrich, "Richard Goldschmidt's 'Heresies' and the Evolutionary Synthesis," *Journal of the History of Biology* 28 (1995): 431–61.

[194] See, for instance, the great call for reform spurred by Vannevar Bush, *Science: The Endless Frontier* (Washington, D.C.: U.S. Government Printing Office, 1945). For the most recent historical account of the development of science during this period, see Arnold Thackray, *Science after '40* (Chicago: University of Chicago Press, 1992). See also Nathan Rheingold, *Science, American Style* (New Brunswick, N.J.: Rutgers University Press, 1991). This call also led to the creation of the National Science Foundation. See Toby Appel, *The History of the National Science Foundation* (forthcoming). See also Benson, Maienschein, and Rainger, eds., *The Expansion of American Biology*.

[195] Initially, members of the SSE balked at officially enrolling their society within the larger category of the AIBS. This resistance was in part due to financial concerns, but also to the fact that the AIBS included experimental biologists who had been denigrating the

any conventional research institution, nor meeting any economic or service-related activity, evolutionary biology as legitimate science would emerge and be sustained because of its unifying properties, which made biology an independent yet unified science within the positivist ordering of knowledge.[196] The unifying principle that biologists in the 1920s had sought now existed as a legitimate science.

No less an authority than Julian Huxley was self-consciously aware that a new science—and a central one at that—had emerged in the second quarter of the twentieth century. Borrowing a term from the title of a book by Arthur Dendy, *Outlines of Evolutionary Biology*,[197] Huxley employed "evolutionary biology" as a substitute for "evolutionary studies" in his *Evolution: The Modern Synthesis*.[198] After the end of the war, and

---

descriptive and nonexperimental sciences. The members of the SSE had much closer ties to the AAAS, to which a large number belonged. The initial move to found what eventually would become the SSE, it should be recalled, took place at the 1939 AAAS meeting. The closer tie to the AAAS justifies the argument that evolutionary biology was supported by the wider audience of scientists, including physicists, chemists, and astronomers, rather than many of the biologists who were close, yet not close enough, to the subject of evolution. See Smocovitis, "Organizing Evolution."

[196] At present there are departments with joint appellations such as "Ecology and Evolutionary Biology" at American universities, and there are numerous centers and programs, but there are no (and have been no) exclusive departments of evolution or evolutionary biology in the United States. See the annual Peterson's *Guide to Graduate Study* for a listing of some thirty-five extant institutions that support their listing under "evolutionary biology" (Peterson's 27th ed., Graduate Programs in Biological and Agricultural Sciences, 1993). Practitioners of evolution reside and have resided in no one locale, and can be found in settings as diverse as universities, museums, and agriculture research institutes. The most frequent connection is with ecology. Evolutionary biology is unlike any of the other disciplines examined by students of science studies. For some of the literature examining the emergence of disciplines, see David Edge and Michael J. Mulkay, *Astronomy Transformed: The Emergence of Radio Astronomy in Britain* (New York: Wiley, 1976); Gerard Lemaine et al., eds., *Perspectives of the Emergence of Scientific Disciplines* (The Hague: Mouton, 1976); Robert E. Kohler, *From Medical Chemistry to Biochemistry: The Making of a Biomedical Discipline* (Cambridge: Cambridge University Press, 1982); Thomas Söderqvist, *The Ecologists: From Merry Naturalists to Saviours of the Nation* (Stockholm: Almqvist and Wiksell International, 1986); Robert Marc Friedman, *Appropriating the Weather: Vilhelm Bjerknes and the Construction of Modern Meteorology* (Ithaca: Cornell University Press, 1989).

[197] The phrase "evolutionary biology" first appears in a passage from naturalist Grant Allen's *Vignettes from Nature*: "and it is these self-same odd, overgrown outer flowers which make the guelder rose so interesting a plant in the eyes of the evolutionary biologist" (Grant Allen, *Vignettes from Nature* [London: Chatto, Winding, and Picadilly, 1881], p. 93). Stephen Jay Gould provided Allen's citation.

[198] In telling the history of what would be the new discipline, Huxley used the phrase "evolutionary studies"—though not strictly in a disciplinary sense—in chap. 1 of *Evolution: The Modern Synthesis* (see p. 23); on p. 31 he explicitly used the phrase "evolutionary biology" in a disciplinary sense. He had been citing Dendy's book in personal notes in the 1930s, and had been actively using the phrase "evolutionary biology" in a disciplinary sense in his publications in the 1920s. English biologists may have adopted his disciplinary sense of the phrase much earlier than their American counterparts. In 1938 Gavin de Beer edited a volume with the title *Essays on Aspects of Evolutionary Biology Presented to E. S. Goodrich* (Oxford: Oxford University Press, 1938). In his 1955 textbook, *Evolution, Genetics and*

with the formation of the Society for the Study of Evolution, the phrase "evolutionary biology" increasingly became an accepted disciplinary appellation.[199] In an address in 1949 Huxley told the following tale of evolutionary biology:

> One of the outstanding events in scientific history has been the emergence, during the second quarter of the present century, of evolutionary biology as a science in its own right. In the phase that followed on Darwin's *Origin of Species* our scientific forebears spoke of the evolution theory, much as in the phase that followed Pasteur, they spoke of the germ-theory of disease. But by the early 20th century, the germ-theory of disease had become swallowed by the science of germs—bacteriology and microbiology—to which it has given rise. In a rather similar way, the evolution theory has today been swallowed in the science of evolution—evolutionary biology. The difference is that, while microbiology is a departmental branch of science, involving a certain definable field, evolutionary biology is a central science, with ideas demarcating all other branches of the life sciences.
>
> This, you may say, is by now a commonplace. However, I do not consider that all the implications of evolutionary biology have been grasped. They have not been fully grasped in the branches of biology: and they have hardly been grasped at all in relation to science as a whole, from physics on the one hand to psychology and the human and social sciences on the other.[200]

In 1953 Huxley once again asserted the existence of the new discipline:

Man (New York: Wiley, 1955), Dobzhansky used "evolutionary biology" consistently in a disciplinary sense.

[199] Both Mayr and Stebbins agree that the phrase gained widespread acceptance shortly after the formation of the SSE. Mayr explicitly used the phrase in a disciplinary sense in a letter to John Aldrich dated 6 August 1947 (Ernst Mayr Papers, Library of the American Philosophical Society, Philadelphia). It was paleontologists—who had come from geology and occupied positions in geology departments, and who could consider themselves evolutionists but could necessarily see themselves as biologists—who found it difficult to adopt the phrase "evolutionary biology." Evolutionary biology was actually included in the subtitle of the journal *Evolution*, which was initially to be called *Evolution: An International Journal of Evolutionary Biology*. In the minutes of the second annual meeting of 1947, this title was emended to *Evolution: An International Journal of Organic Evolution*. I could find no reason for this change in title in any of the documents I examined in the archives of the American Philosophical Society, but it is likely that the paleontologists resisted the widespread application of the term. Simpson repeatedly used the terms "evolutionary studies," "evolution," or "paleontology" as well as making the distinction between paleontology and "neo-biology." See the discussion below on the emergence of "paleobiology."

[200] "Evolution and Scientific Reality," manuscript dated 1949, Box 67.7, Huxley Papers.

Evolutionary science is a discipline or subject in its own right. But it is the joint product of a number of separate branches of study and learning. Biology provides its central and largest component, but it has also received indispensable contributions from pure physics and chemistry, cosmogony and geology among the natural sciences, and among human studies from history and social science, archaeology and prehistory, psychology and anthropology. As a result, the present is the first period in which we have been able to grasp that the universe is a process in time and to get a first glimpse of our true relation with it. We can see ourselves as history, and can see that history in its proper relation with the history of the universe as a whole.[201]

Though he was self-aware—indeed, played an important role in the emergence of evolutionary biology—Huxley underestimated the sensitivity of his wide audience of readers and the eagerness with which they had sought to unify all the sciences within a coherent worldview. Physicists (who themselves had set the standards for the quest for grand unified cosmic theories), chemists, astronomers, and social scientists—from Sir James Jeans, Arthur Eddington, Fred Hoyle, and Jacob Bronowski to Loren Eiseley—did in fact rapidly adopt versions of the progressive evolutionary framework that drew on evolution by means of natural selection.[202] The end result would be the development of an evolutionary cos-

[201] Julian Huxley, *Evolution in Action* (New York: Harper and Brothers, 1958), pp. 1–3. The celebration in honor of the fiftieth anniversary of the rediscovery of Mendel's genetics at the "Golden Jubilee of Genetics" held at Ohio State University in 1950 gave Huxley once again the opportunity to argue for the special position of evolutionary biology: "Evolutionary biology is the best example of the historical approach in science, as opposed to the observational or the experimental; and thus it provides the best link between physicochemical science and human history. Furthermore, it is the only bridge between the purely material world of lifeless matter and the human world of mind, thus enabling every aspect of reality to be seen as part of one comprehensive process" (Huxley, "Genetics, Evolution and Human Destiny," p. 612).

[202] Max Dresden and Pierre Noyes (fairly representative of their generation of physicists) at the Stanford Linear Accelerator shared their historical perspectives of the development of physics in the interwar and postwar periods. Both had closely followed developments in evolutionary biology. Both had read *Evolution: The Modern Synthesis* and other essays by Huxley, as well as the semipopular works of the Haldanes. One physical scientist (trained in mathematics) who also drew heavily on Huxley's framework was Jacob Bronowski. Huxley and Bronowski collaborated on writing volumes of the Macdonald Illustrated Library in the 1963–65 period. Bronowski's televised "Ascent of Man" series echoed Huxley's evolutionary humanism. Adopting both Huxley's and Bronowski's evolutionary humanism, Carl Sagan televised this evolutionary philosophy through his "Cosmos" series. Episode 2, entitled "One Voice in the Cosmic Fugue," laid the groundwork for the series (episode 1 was an introductory synopsis of the thirteen-part series). It introduced evolution by means of natural selection. Sagan demonstrated the efficacy of natural selection through the example of the Heike crab—an example also used by Huxley in his essay "Life's Improbable Likenesses," chap. in *New Bottles for New Wine* (London: Chatto and Windus, 1957), pp. 137–

mology—with cosmological change demonstrating the same law-like regularities with clearly defined mechanical causes of change and material entities on which they acted.[203] Cosmic, galactic, stellar, planetary, chemical, organic evolution, and cultural evolution emerged as a continuum in a "unified" evolutionary cosmology by the 1940s.[204] Inheriting this evolutionary cosmology, a charismatic young University of Chicago-based National Science Foundation predoctoral fellow in physics and astronomy, studying with planetary scientist Gerard P. Kuiper, chemist Harold Urey, Sewall Wright, and Hermann J. Muller, published his first article in *Evolution*. In 1957, in an article entitled "Radiation and the Origin of the Gene," Carl Sagan, representing the convergence and linkage of these scientific disciplines grounded in the modern synthesis of evolution, made his scientific debut.[205]

Even those who stressed the persistent problems that remained in disciplines not well integrated in the modern synthesis gave generous credit to the new evolution grounded in genetics. Though he pointed to persistent problems in embryology and evolution and echoed his embryological brethren like Goldschimdt and Schmalhausen in criticizing the inadequacies of mere mathematical genetics in his 1953 "Epigenetics and

---

54. The Heike crab had been a favorite with Hermann J. Muller, and both Huxley and Sagan had close ties to Muller. In the early 1980s Huxley's form of evolutionary humanism—arguably not significantly altered—was therefore to be transmitted through the latest technology to one of the largest popular science audiences of all time: 16 million viewers.

[203] A special issue of *Scientific American* entitled "The Universe" published in 1956 surveyed current developments in cosmology. See especially George Gamow's article, "The Evolutionary Universe," *Scientific American* 195 (1956): 136–54. On the evolution of the galaxies, see the contribution by Jan H. Oort, "The Evolution of Galaxies," pp. 100–108. But not all cosmologists supported this model. See the contribution by Fred Hoyle, "The Steady-State Universe," pp. 157–66. It will be recalled that one of the strongest backers of evolution had been the astronomer Harlow Shapley. With his support, Mayr and Simpson were able to obtain start-up funds from the American Philosophical Society to found their society—while embryologists and physiologists like E. G. Conklin vetoed proposals to establish an official society for the study of evolution.

[204] Huxley stated in his *Evolution in Action* that some new term was needed for the whole or comprehensive process. He decided that while philosophers of science might wish to coin a new term, he would use "evolution." The overall process of evolution for Huxley consisted of three main phases: inorganic (or cosmological), organic (or biological), and the human (psychosocial). Though they had their differences, all three were sectors of a universal process (Huxley, *Evolution in Action*, pp. 2–3). With the "discovery" of Precambrian microfossils in 1954 through the imaging technology of the electron microscope, Elso S. Barghoorn and Stanley A. Tyler introduced cellular evolution into the continuum (this was later extended by Lynn Margulis). Biochemical and molecular evolution were introduced in the 1960s by biochemists like Richard Dickerson. Each of these evolution communities represents a diverse and heterogeneous set of practices unified by the evolutionary narrative.

[205] Carl Sagan, "Radiation and the Origin of the Gene," *Evolution* 11 (1957): 40–55. Sagan had been a student at the University of Chicago. He had also worked with Joshua Lederberg.

Evolution,"[206] Conrad H. Waddington could still feel that the "advances made" in evolution in the past thirty years were so "striking that they may even be taken to have reached their goal with some degree of finality."[207] But for Waddington, as for subsequent embryologists (soon-to-turn developmental biologists), the final synthesis between embryology and evolution had to await further developments in biochemical and physiological genetics.

Waddington's call for further synthesis (which formed the bulk of his lifework) was echoed by J. B. S. Haldane in the foreword to the same volume. Haldane assessed the status of the modern synthesis of evolution as being in its own developmental "instar"; the synthesis was not yet ready for "a new moult," though he added that "signs of new organs are perhaps visible."[208] What neither of these individuals could at that time see was the success of the new evolutionary science: the volume that carried Waddington's supposed "critique" of the modern synthesis had by the mid-1950s rendered evolution as an accepted and indeed integral part of experimental biology. The very society that Julian Huxley had helped found, the Society for Experimental Biology, devoted an entire symposium to the subject in 1952 in collaboration with the Genetical Society at Oxford, and then published the proceedings in an issue devoted to "Evolution"—the very subject that had been deemed a nonexperimental nonscience thirty years earlier. Included herein were the widest possible range of biological topics representing the biological sciences: the origins of life, biochemistry, population genetics, cytology, immunology, embryology, and social behavior. These studies drew on the diversity of life with exemplars from *Escherichia coli*, to plants like *Primula*, to newts, to butterflies, and to vertebrates and primates; the

[206] Waddington stated: "It has been primarily those biologists with an embryological background who have continued to pose questions: Goldschmidt with his ideas of the 'unbridgeable gap' and systemic mutations; Schmalhausen with his notion of stabilizing selection; and Dalcq, whose latest article (1951) has the slightly ironical title 'Le problème de l'évolution est-il près de'être résolu?'" (Conrad H. Waddington, "Epigenetics and Evolution," *Evolution: Symposia of the Society for Experimental Biology* 7 [New York: Academic Press, 1953], p. 186).

[207] Waddington seemed especially unhappy with the suggestion that the mathematical population geneticists had resolved all the problems of evolution. Waddington made the goal for a true synthesis between embryology and evolution his lifework. For some of his original work, see Conrad H. Waddington, *The Strategy of the Genes* (London: Allen and Unwin, 1957); for an autobiographical account, see *The Evolution of an Evolutionist* (Ithaca: Cornell University Press, 1975). For historical analysis of Waddington's goal of the synthesis between embryology and evolution, see Scott Gilbert, "Epigenetic Landscaping: Waddington's Use of Cell Fate Bifurcation Diagrams," *Biology and Philosophy* 6 (1991): 135–54.

[208] J. B. S. Haldane, "Foreword," *Evolution: Symposia of the Society for Experimental Biology* 7 (New York: Academic Press, 1953), p.xviii–xix.

topic of "evolution" had appeared to unify the diversity of biological sciences!

Historians of science in the United States gearing up to assess the critical problems in their young field could not help but take note of the widespread talk on the maturation of biology and the "unifying properties" of the science of evolution then taking place. From this mid-1950s presentist vantage point, historians of science regarded Darwin's theory of evolution as the unifying principle of the biological sciences that had matured in the nineteenth century. For this reason, organizers of a conference singled out what was clearly an important issue in the 1950s: "Evolution as a Unifying Principle in the Biological Sciences of the Nineteenth Century." Assigned this topic, J. Walter Wilson could not help but note the contributions of Pasteur and Virchow to the larger nineteenth-century biological sciences; he altered his assigned topic to accommodate their contributions under the title "Biology Attains Maturity in the Nineteenth Century." Interestingly, Conway Zirkle's commentary to Wilson's article reassessed the contributions of Pasteur and Virchow more critically, stressing their eliminative or negatory influence on the emergence of biology. Returning to the contributions of Darwin and evolutionary theory, Zirkle extended "the unifying principle" of evolution to even more sciences in light of the 1950s experiments on the origin of life, promoting and extending his contemporary belief far beyond that possible in Darwin's time. Understanding the historical conditions for the emergence of the biological sciences thus became one of the *Critical Problems in the History of Science* (the title of the final edited volume).[209]

Stronger validation of the belief in the existence of a unified biological science and its secure location would come from philosophers who reexamined the foundations of biology in the 1950s. Writing within the vehicle founded by the Unity of Science Movement to encompass all knowledge, *Foundations of the Unity of Science*, Felix Mainx, who can be viewed as occupying a philosophical niche not unlike that of Woodger and other philosophical biologists in the 1920s, had a perspective on the status of the twentieth-century biological sciences vastly more confident than that of his predecessors, who had examined the status of biology as a science.[210] Rather than hurling criticism after criticism at a disunified and fragmented infant science, Mainx could instead tell the tale of the

---

[209] See Marshall Clagett, ed., *Critical Problems in the History of Science* (Madison: University of Wisconsin Press, 1959). See specifically the contributions by J. Walter Wilson, John C. Greene, Richard H. Shryock, and Conway Zirkle.

[210] J. H. Woodger served as translator and critic for Mainx's monograph.

science of biology, empirical, grounded, and unified. Though the central problematic of biology, evolution, was by no means completely understood nor even necessarily testable (by Mainx's philosophical standards),[211] questions raised by considerations of evolution had performed an "invaluable heuristic service in all branches of biology."[212] What had formerly been mistaken as disparate points of view had undergone a synthesis, reconciling differences of opinion and conflicts and leading to the construction of a coherent and consistent evolutionary worldview:

> In recent years a clear and far-reaching approximation of the various viewpoints has taken place, and various books, as well as discussions at congresses and symposia, allow us to recognize clearly the development of a new synthesis of all possible points of view. Moreover, in this most difficult branch of biological investigation a phenomenon has become clear which in many other branches of biology, and in all pure empirical sciences, can be regarded as a touchstone for the fundamental confirmation of the methodical path of these sciences: the spontaneous convergence of all lines of development in the science toward a closed, consistent picture of the world.[213]

Mainx concluded with the following thoughts on biological science:

> As an unavoidable consequence of its rich development, biology has experienced an especially marked subdivision into special branches,

[211] Other philosophers would continue to view evolution as a highly problematic science. See Marjorie Grene, "Two Evolutionary Theories," *British Journal for the Philosophy of Science* 9 (1959): 110–27, 185–93; and Karl Popper, *Objective Knowledge* (Oxford: Clarendon, 1972). Both were to modify their initially critical positions. One philosopher who viewed evolution on favorable scientific terms was Ernst Cassirer. See *The Problem of Knowledge: Philosophy, Science, and History since Hegel*, trans. William H. Woglom and Charles W. Hendel (New Haven: Yale University Press, 1950). Mayr responded to each of these philosophers.

[212] Felix Mainx, *Foundations of Biology*, in *Foundations of the Unity of Science: Toward an International Encyclopedia of Unified Science*, 1, no. 9 (Chicago: University of Chicago Press, 1955), p. 52. The exact quotation reads:

Although the whole complex of problems thrown up by the theory of evolution must in many of its parts always remain in the stage of a hypothesis which is not testable in practice, yet it has done invaluable heuristic service in all branches of biology and has therefore become an indispensable part of the method of biology. In view of the multiplicity of points of view regarding evolutionary questions indicated above, it is not surprising that representatives of the various subdivisions of biology, such as systematists, morphologists, paleontologists, biogeographers, and geneticists often put the problems differently and give the hypothesis a different meaning from their several points of view or estimate their empirical confirmation differently. Differences of opinion which come to light in this way have often originated clarifying discussions and so led to fruitful new efforts. Unfortunately, such conflicts are often unpleasant and fruitless, owing to a lack of understanding of the logic of science.

[213] Ibid., p. 52.

and this carries with it a certain danger of onesidedness. The synthesis of the results of biology nevertheless goes on throughout consistently and fruitfully and leads to a constant development of the science. There is in biology no 'crisis,' as has sometimes unjustly been stated. The synthesis of the results of biology with those of the remaining natural sciences has been fruitfully established in many borderline regions and leads to an empirical world picture which is on the whole consistent and unified, if incomplete.

A critical study of foundations and methods which could only be hinted at in this monograph would certainly be very useful in biology. But here biology occupies no special position, because such problems are common to all of the empirical sciences.[214]

Thus, by 1955—the same year that witnessed the death of Albert Einstein—a group of "super-Einsteins" could argue for a unified science of biology.[215] Not only had the discipline long thought to be disunified, in its infancy, and rife with metaphysics become a unified science, but now, too, it was an empirical, mature science, secure in its foundations and well positioned within the positivist ordering of knowledge—intermediate between the physical sciences and the social sciences. Evolution, stretching from the gene to the human to human culture, would bind and link the mechanistic and materialistic frameworks with the human sciences.[216] Reducible to the physical world of the gene and grounded in the fundamental mathematical principles of population genetics, the disciplines within the positivist ordering of knowledge stood independent yet united. Biology would not be any more or less outstanding or problematic than any of the physical sciences. Biology, now enough of an axiomatic science with its own logical principles, was no longer in its metaphysical stage of development but had become a mature science that could rival physics and chemistry. Biology—indisputable now as *the science of life*—would also make possible a more meaningful existence for "modern man." The "modern synthesis of evolution" thus lent unity to the mature science of biology, but also unity to the other disciplines of science within a coherent worldview; evolutionary biology emerged as the central balancing point of the unified sciences. In the preface to the first edition of the Pelican Biology series, *The Theory of Evolution* (1958),

[214] Ibid., pp. 84–85.
[215] It will be recalled that William Morton Wheeler earlier noted that it would take "a few super-Einsteins to unify biology."
[216] Sociobiology was to emerge from the continuum between genetics and the social sciences. See the discussion below.

John Maynard Smith could therefore write: "The main unifying idea in biology is Darwin's theory of evolution through natural selection."[217] In this very same year Edward O. Wilson—inheriting the synthesis—taught the first course at Harvard University initiating practitioners into the emerging central science, Biology 144: *Evolutionary Biology*. Finally, in 1962 a university textbook entitled "*A Synthesis of*" *Evolutionary Theory* appeared, fulfilling the ultimate goal begun with the origins of "science" in antiquity. Situating the quest for cosmic unity in ancient Ionia, Herbert J. Ross concluded: "This full sweep of evolutionary theory is indeed tantamount to a unifying principle of the universe."[218] But, by then, Dobzhansky's 1955 undergraduate textbook embodying, in suitable order, his life-interests, *Genetics, Evolution and Man*, was taking this message—replete with history and philosophy—to an emerging young generation who inherited the appellation of "evolutionary biologists".[219]

The maturation of biology and the emergence of evolutionary biology took place just as the centenary of the publication of Darwin's *Origin* was approaching. Gathering to reexamine and reassess the work of this "great man of science," evolutionary biologists and historians would begin to contribute to the burgeoning literature of Darwin studies.[220] While cele-

[217] John Maynard Smith, "Preface to the first edition" of *The Theory of Evolution* (Harmondsworth, Middlesex: Penguin, 1958; rev. and repr. in the 3d ed. 1975; repr. 1977).

[218] Herbert J. Ross, "*A Synthesis of*" *Evolutionary Theory* (Englewood Cliffs, N.J.: Prentice Hall, 1962), p. vii. Ross also wrote: "Perhaps no field of study combines so many facts and theories from so broad a spectrum of science as does the study of evolution. It is a synthesis of facts and theories from zoology, botany, biochemistry, genetics, geology, ecology, and many other bordering fields." He continued, stressing the continuum of evolutionary processes: "Such a philosophy of evolution, embracing the entire scope of the evolutionary process, starts with pre-stellar evolution and continues by logical steps to the evolution of biomes" (quotation on pp. vii–viii).

[219] Subsequent editions of Dobzhansky's *Genetics and the Origin of Species* stressed the unified new science of evolutionary biology. The third edition of 1951 began with the following: "Finally, the most recent developments indicate a trend toward synthesis of what were often diverging historical and causal approaches, and toward emergence of a unified evolutionary biology" (Theodosius Dobzhansky, *Genetics and the Origin of Species*, 3d ed. [New York: Columbia University Press, 1951], pp. 11–12).

[220] The fourteenth annual meeting of the SSE was held in conjunction with the University of Chicago's Darwin Centennial Celebration in honor of the one hundreth anniversary of the publication of Darwin's *Origin* on 24–28 November 1959; see the three volumes edited by Sol Tax (especially vol. 3) for an account of the festivities, which included commemorative ceremonies with participants in full academic regalia (Sol Tax and Charles Callender, eds., *Evolution after Darwin*, vol. 3 [Chicago: University of Chicago Press, 1960]). See also Vassiliki Betty Smocovitis, "Celebrating Darwin," (forthcoming). The new literature included Loren Eiseley, *Darwin's Century: Evolution and the Men Who Discovered It* (New York: Doubleday, 1958); John C. Greene, *The Death of Adam: Evolution and Its Impact on Western Thought* (Ames: Iowa State University Press, 1959); C. D. Darlington, *Darwin's Place in History* (Oxford: Blackwell, 1959); Gertrude Himmelfarb, *Darwin and the Darwinian Revolution* (Garden City, N.Y.: Doubleday, 1959); Gavin de Beer, *Charles Darwin: Evolution by Natural Selection* (Edinburgh: Nelson, 1963). Scientists were to favor de Beer's reading of Darwin; Himmelfarb was not to fare as well.

brants revelled in the one-hundredth birth-year of Darwin's "Great Work," they also took the occasion to reassess their own evolutionary state of the art. Thus, on the exact anniversary date of 24 November 1959, the architects of the new synthesis joined with evolutionists from all over the world to rejoice in the new synthesis of evolution at the grandest of all celebratory rites, including a convocation ceremony, discussion panels, and a musical play based on the life of Darwin entitled *Time Will Tell*. Reenacting Darwin's "Voyage of the Beagle" and rereading the present into the past,[221] they reinvented Darwin and Darwinism as *Neo*-Darwinism, and reinterpreted his "theory of descent with modification" as evolution by means of natural selection.[222] Darwin was to be reconstructed once again as the "founding father" of the new discipline of evolutionary biology.[223] Yet though Darwin was to be repeatedly hailed as the Newton of biology,[224] it was the "modern synthesis" that would function as the biological analogue of the "Newtonian synthesis" in the grand narrative of the history of science.[225]

[221] This phrase is borrowed from Galison's historiographic article on Maxwell; see Peter Galison, "Re-Reading the Past from the End of Physics: Maxwell's Equations in Retrospect," in Loren Graham, Wolf Lepenies, and Peter Weingart, eds., *Functions and Uses of Disciplinary Histories*, (Dordrecht: D. Reidel, 1983), pp. 35–51.

[222] See, for instance, the volume of selections from Darwin and Wallace compiled by Gavin de Beer, *Evolution by Natural Selection* (Cambridge: Cambridge University Press, 1958); and see de Beer's biography of Darwin, *Charles Darwin: Evolution by Natural Selection* (Edinburgh: Nelson, 1963).

[223] "Founding father" stories emerge from, and sustain, disciplinary identities. The identity of the founding father is altered and permuted as it is reconstituted with each telling of the story. This disciplinary interpretation resolves the problems introduced by Jan Sapp in "The Nine Lives of Mendel," in H. E. Legrand, ed., *Experimental Inquiries* (Dordrecht: Kluwer, 1990); Jan Sapp, *Where the Truth Lies* (Cambridge: Cambridge University Press, 1990). Darwin has had—and will have—many, many lives.

[224] Julian Huxley reaffirmed Darwin as the "Newton of Biology" in an essay prepared for the University of Chicago Darwin Centennial Celebration: see "The Emergence of Darwinism," in Sol Tax, ed., *Evolution after Darwin*, vol. 1, *The Evolution of Life* (Chicago: University of Chicago Press, 1960), pp. 1–21. According to Huxley it was Alfred Russel Wallace who first called Darwin "the Newton of Natural History." Darwin is still frequently referred to as the "Newton" of biology. Philip Kitcher used this phrase on p. 54 of *Abusing Science: The Case against Creationism* (Cambridge: MIT Press, 1982). More recently, Mayr has reaffirmed this view of Darwin: see Ernst Mayr, "The Ideological Resistance to Darwin's Theory of Natural Selection," *Proceedings of the American Philosophical Society* 135 (1991): 123–39.

[225] It may be argued that the architects of the evolutionary synthesis function as the analogues of Newton; the architects would possibly not dispute this view. Julian Huxley opened his Jubilee of Genetics contribution with the following assertion: "In celebrating the jubilee of genetics, we celebrate at the same time the culmination and the triumph of the biological revolution inaugurated during the nineteenth century, which is destined, I believe, to have even more profound results than the physical revolution inaugurated during the seventeenth" (Huxley, "Genetics, Evolution and Human Destiny," p. 591). The argument that the Newtonian synthesis and the evolutionary synthesis—especially after sociobiology—bear some resemblance to each other has been made by Gerald Holton: see "Analysis and Synthesis as Methodological Themata," chap. in *The Scientific Imagination: Case Studies* (Cambridge: Cambridge University Press, 1978), pp. 111–51.

THE UNITY OF LIFE AND THE DIVERSITY OF LIFE: BIOLOGICAL AUTONOMY IN THE POST-*Sputnik* BIOLOGICAL SCIENCES—A POSTSCRIPT

While the struggle to unify the biological sciences appeared to be over by 1955, the struggle to preserve and maintain the unity, autonomy, and location of the biological sciences would continue. The "architects"—a self-designated term—of the modern synthesis would function as the "unifiers" of the biological sciences. Redefining as they renegotiated disciplinary boundaries, the architects were to preserve the delicate balance between unity and autonomy in the biological sciences.

With the launching of *Sputnik* and the reorganization of federal research, American science increasingly came to represent a greater heterogeneity of biological practices. Debates on the relationship between the newer biological sciences flared up repeatedly throughout the 1960s, with discussions focusing on the location of the various fields of the biological sciences on the "totem pole" of science.[226]

One of the newer biological sciences to emerge at this time had sprung out of the burgeoning U.S. space program. Drawing heavily on biochemistry, investigators of the new "exobiology"—Joshua Lederberg's new alternative to earth or ground-based "esobiology"—began to examine closely the biochemical basis for the origins of life on earth and on other worlds. Grounded in the theoretical foundations of the modern synthesis of evolution with its well-defined material entities and mechanical causes of change, and assuming the linkage between evolutionary progress and technological progress, the "Drake Equation"[227]—formu-

---

[226] See Charles C. Davis, "Letters: Biology Is Not a Totem Pole," *Science* 141 (1963): 308–10.

[227] There are many versions of the Drake Equation. Each of the factors in the equation represent an entity that emerged from evolutionary cosmology: the universe, galaxies, solar systems, suns, planets, molecules. The right combination led probabilistically to intelligent and technological forms of life. The following is one variation on the theme of the Drake Equation that found its way to 16 million viewers of "Cosmos" when it was first broadcast in 1980:

Where N_*, the number of stars in the Milky Way Galaxy; f_p, the fraction of stars that have planetary systems; n_e, the number of plants in a given system that are ecologically suitable for life; f_l, the fraction of otherwise suitable planets on which life actually arises; f_i, the fraction of inhabited planets on which an intelligent form of life evolves; f_c, the fraction of planets inhabited by intelligent beings on which a communicative technical civilization develops; and f_L, the fraction of a planetary lifetime graced by a technical civilization. Written out, the equation reads: $N = N_* f_p n_e f_l f_i f_c f_L$. All the f's are fractions, having values between 0 and 1; they will pare down to the large value of N. (Carl Sagan, *Cosmos* [New York: Random House, 1980], p. 299)

lated by radioastronomer Frank Drake[228]—argued for the probability of the existence of technologically intelligent forms of extraterrestrial life. With their organized search for extraterrestrial intelligence (SETI),[229] exobiologists were able to capture both the scientific and popular limelight in the early 1960s. But for "esobiologists" the biochemical basis for the origins of life, which stressed heavily the unity of life based on a common biochemistry, threatened to lead to complete reduction and hence engulfment by the physical sciences. Disciplining to preserve the delicate balance once again, Simpson entered the exobiology circle to construct his celebrated argument for "the nonprevalance of humanoids."[230] Evoking historical contingency and the role of chance in evolutionary events, he successfully negotiated the boundaries of evolutionary biology, giving enough autonomy to the biological sciences. In so doing he also reintroduced the issue of historical contingency back into evolutionary biology.[231]

But by far the greatest danger to the preservation of this balance was

---

For an illuminating personal account of the derivation of the equation at the famed Greenbank conference and the history of SETI, see Frank Drake and Dava Sobel, *Is Anyone out There? The Scientific Search for Extraterrestrial Intelligence* (New York: Delacorte, 1992).

[228] For much of his life, Drake had been associated with a very active group of exobiologists at Cornell University. For an interesting sociological portrait of the SETI members, see David W. Swift, *Seti Pioneers: Scientists Talk about Their Search for Extraterrestrial Intelligence* (Tucson: University of Arizona Press, 1990). According to Swift, SETI activity seemed to flourish in two locations: Cornell University and the Palo Alto area. The present location of the organization of the SETI Institute is in Mountain View, California.

[229] Originally the group met under the rubric CETI (Communication with ExtraTerrestrial Intelligence). American CETI conferences began officially in 1961 under the auspices of the National Academy of Sciences; see Carl E. Sagan, ed., *Communication with Extraterrestrial Intelligence* (Cambridge: MIT Press, 1973), for the conference proceedings of the international meetings held in Byurakan, Soviet Armenia, in 1971.

[230] George Gaylord Simpson, "The Nonprevalance of Humanoids," *Science* 143 (1964): 769–75. See also Simpson's exceedingly vitriolic "Added Comments on the 'Nonprevalance of Humanoids,'" in *Communication with Extraterrestrial Intelligence*, pp. 362–64.

[231] The first generation of evolutionary biologists (with some exceptions) did not strongly advocate the SETI program. See Edward Regis, Jr., ed., *Extraterrestrials: Science and Alien Intelligence* (Cambridge: Cambridge University Press, 1985). For confirmation of this position, see Ernst Mayr's contributed essay, "The Probability of Extraterrestrial Intelligent Life," which constructs an argument similar to Simpson's. More recent evolutionary biologists have supported SETI research, at least on paper: Thomas Eisner, Stephen J. Gould, David Raup, Dale A. Russell, and E. O. Wilson were included as signatories among seventy-odd distinguished scientists who signed a petition to support SETI that appeared in *Science*, 29 October 1982. This had followed on the heels of the spectacular success of "Cosmos" in 1980. See Drake and Sobel, *Is Anyone Out There?* for the discussion and the list and titles of the signatories. More recently, Ernst Mayr launched an attack in *Science* against funding for SETI research. Ernst Mayr, "Letters: The Search for Intelligence," *Science* 260 (1993): 1522–23. See the responses in defense of the search by Frank Drake, John D. Rummel, and David Raup: "Letters: Extraterrestrial Intelligence," *Science* 260 (1993): 474–75.

to come from the determination of the biochemical and molecular basis of life. Two scientific events in 1953 were to lead to the momentary destabilization of the biological sciences: the experiments of Stanley Miller and Harold Urey confirmed the biochemical origins of life;[232] and James Watson's and Francis Crick's efforts led to the articulation of the structure of the macromolecular "secret of life"—DNA.[233] As research in molecular biology and biochemistry intensified, these two new disciplines became heavily supported and institutionalized.[234] With these two disciplines emerging, the linkages among physicists, chemists, and biologists solidified further. With the subsequent articulation and refinement of the molecular basis for genetic change, biology faced its greatest threat of complete engulfment by the physical sciences.[235]

The beginnings of the split between organismic and molecular biology were felt in the 1950s at Harvard, where Ernst Mayr defended organismic biology against molecular biology in a celebrated exchange with biochemist George Wald.[236] In 1959 Ernst Mayr faced head on the same audience that he viewed responsible for the decline in the status of evolutionary biology at the Cold Spring Harbor Symposium on Quantitative Biology. Inspired by Waddingon's earlier comments on the synthesis, Mayr charged that mathematical population geneticists had simplistically reduced evolution to the action of mere gene frequencies, which in turn ignored interactive gene effects, the organism itself, and the organism's natural environment. He made an impassioned plea for these "bean-bag" geneticists to reconsider the complex problems remaining to be solved by organismic and evolutionary biologists. This was especially critical with the rise of molecular biology, which was rendering organismic and evolu-

[232] Stanley Miller, "A Production of Amino Acids under Possible Primitive Earth Conditions," *Science* 117 (1953): 528–29.

[233] Gunther Stent, ed., *The Double Helix* (New York: W. W. Norton, 1980). Available histories of the early years of molecular biology are numerous: Robert Olby, *The Path to the Double Helix* (Seattle: University of Washington Press, 1974); Franklin H. Portugal and Jack S. Cohen, *A Century of DNA: A History of the Discovery of the Structure and Function of the Genetic Substance* (Cambridge: MIT Press, 1977); and the now-classic, Horace Freeland Judson, *The Eighth Day of Creation: Makers of the Revolution in Biology* (New York: Simon and Schuster, 1979). No history of the discovery of the structure of DNA is complete without taking into account the work of Rosalind Franklin: see Anne Sayer, *Rosalind Franklin and DNA* (New York: W. W. Norton, 1975).

[234] See Robert Kohler, *From Medical Chemistry to Biochemistry: The Making of a Biomedical Discipline* (Cambridge: Cambridge University Press, 1982).

[235] For a historical discussion of what many identified as the "DNA bandwagon effect" and how it fueled antireductionist philosophical arguments, see John Beatty, "Evolutionary Anti-Reductionism: Historical Reflections," *Biology and Philosophy* 5 (1990): 199–210.

[236] E. O. Wilson includes a chapter entitled "Molecular Wars" describing the interactions at Harvard between the various biologists. E. O. Wilson, *Naturalist* (Washington, D.C.: Island Press/Shearwater Books, 1994).

tionary biology obsolete. Linking the future survival of man in the competitive, technologically challenging global community with a study of its evolution, Mayr thus gave equal ranking to evolutionary biology:

> We live in an age that places great value on molecular biology. Let me emphasize the equal importance of evolutionary biology. The very survival of man on this globe may depend on a correct understanding of the evolutionary forces and their application to man. The meaning of race, of the impact of mutation, whether spontaneous or radiation-induced, of hybridization, of competition—all these evolutionary phenomena are of the utmost importance for the human species. Fortunately the large number of biologists who continue to cultivate the evolutionary vineyard is an indication of how many biologists realize this: we must acquire an understanding of the operation of the various factors of evolution not only for the sake of understanding our universe, but indeed very directly for the sake of the future of man.[237]

As the "classical" sciences were endangered by the popularity—and financial support—of these "newer" sciences, Ernst Mayr wrote for a wide audience of scientists to create space for both.[238] In the early 1960s the architects and unifiers were to address the ensuing "crisis"—once again—in the biological sciences.[239]

Warding off reduction to the physical sciences became a primary concern for Ernst Mayr, the architect most sensitive to philosophy at this time. It was because he himself had sympathized (initially at least) with the logical positivists and had advocated the unity of science that Mayr was the first to be alerted to the danger of the physicalist reduction that accompanied the unification of the sciences along with the rise of biochemistry and molecular biology.[240] With the combination of the increasing emphasis placed on population and "bean-bag" genetics in the 1950s,

[237] Ernst Mayr, "Where Are We?" *Cold Spring Harbor Symposia on Quantitative Biology* 24 (1959): 1–14.

[238] Ernst Mayr, "The New versus the Classical in Science," *Science* 141 (1963): 765. Mayr wrote: "Far more important, for the general well-being of American science and the attainment of a healthier balance between classical and frontier fields, is more financial and moral support for the classical areas." A somewhat different point was made in a letter to the editor of *Science* later in 1963 that strongly supported "Modern Biology" over "Traditional Biology." See Fred E. Hahn, "On Traditional and Modern Biology," *Science* 141 (1963): 1240–42.

[239] See George Gaylord Simpson, "The Crisis in Biology," *The American Scholar* 36 (1966–67): 363–77; Theodosius Dobzhansky, "On Cartesian and Darwinian Aspects of Biology," *The Graduate Journal* 8 (1968): 99–117.

[240] Ernst Mayr had been aware of the Unity of Science Movement but rejected it when he realized that the unity of science would take place by the reduction of biology to physics (letter to author, 8 October 1991).

and the popularity of the physicalist philosophy of Ernest Nagel, the unification of biology threatened to lead to a complete reduction to the physical sciences. Writing for the wider audience of scientists in 1961, Mayr constructed a powerful argument supporting the autonomous yet unified status of the biological sciences.[241] Comparing causality in biology and physics, he argued that "causality in biology" was a "far cry from causality in classical mechanics." The structure of biological science was far more complex and had its own emergent properties unlike any of the physical sciences. Biology—from then on—would consist of two biologies: functional biology and evolutionary biology. While materialistic and mechanistic principles accounted for functional biology, properties in some manner emergent yet well within legitimate science were to account for evolutionary biology. Evolutionary biology—the biology of ultimate, not proximate, causes—would therefore "lift" biology, through an argument for "emergence," from complete reduction to the physical sciences, and at the same time would become the unifying element of a unified biology. In so arguing, Mayr was beginning to ground a new philosophy of science, based not solely on physics, but on biology! In 1963 Simpson picked up and extended Mayr's argument for two biologies not only to argue against reduction to the physical sciences, but also to argue for the centrality of biology in the drive to unify the sciences. He wrote the following for a wide audience: "Biology, then, is the science that stands at the center of all science. It is the science most directly aimed at science's major goal and most definitive of that goal. And it is here, in the field where all the principles of all the sciences are embodied, that science can truly become unified."[242] Then comparing biology—clearly the more complicated science—to physics, Simpson argued forcefully that biology was actually more "scientific" than even physics, long held as the exemplar science. He wrote:

the life sciences are not only much more complicated than the physical sciences, they are also much broader in significance, and they penetrate much farther into the exploration of the universe that *is* science than do the physical sciences. They require and embrace the data and *all* the explanatory principles of the physical sciences and then go far beyond that to embody many other data and additional explanatory principles that are no less—that are, in a sense, even more—scientific.[243]

---

[241] Ernst Mayr, "Cause and Effect in Biology," *Science* 134 (1961): 1501–6.
[242] George Gaylord Simpson, "Biology and the Nature of Science," *Science* 139 (1963): 81–88; reprinted in *This View of Life*. Quotation on p. 107.
[243] Ibid., p. 104.

But the danger of complete reduction to molecular biology—and ultimately physics—was avoided, indeed made to work *for* biology, through the unifying argument articulated by Dobzhansky. Stretching the continuum from the gene to the human, Dobzhansky incorporated the molecule as the new level of evolution. This new level would serve integrative functions accounting for both the unity of life, with its connection to the physical sciences, and the diversity of life, with its connection to organismic biology. Physics, chemistry, and molecular biology accounted for the *unity* of life, whereas organismic biology, ecology, and the social sciences accounted for the *diversity* of life. Evolutionary mechanisms were responsible for the unity and diversity of life, so that the unity of the sciences was preserved.

> The world of life can be studied from two points of view—that of its unity and that of its diversity. All living things, from viruses to men, have basic similarities. And yet there is an apparently endless variety of living beings. Knowledge and understanding of both the unity and the diversity are useful to man. I like, however, to stress here not the pragmatic aspect, not the applied biology, but the aesthetic appeal. Both the unity and the diversity of life are fascinating. Some biologists find the unity more inspiring, others are enthralled by the diversity. This is evidently a matter of personal taste, and a classical adage counsels that tastes are not fit subjects for disputation (although this is what most disputations are about). The consequence of the polymorphism of tastes is that there always will be different kinds of biologists and different subdivisions of biology. Some of the subdivisions may be offering more fleshpots than others, and hence will be more popular, especially among those for whom the fleshpots are the prime considerations. Other subdivisions will, however, continue to attract some votaries.

But it was the organismic biological sciences—situated between the molecular and physical sciences and the social sciences—that would begin to answer the "big question" facing "every generation" that "must solve it in relation to the situation it faces: What is "Man?"[244] Extending

---

[244] See Theodosius Dobzhansky, "Biology, Molecular and Organismic," *American Zoologist* 4 (1964): 443–52; quotations on pp. 448–51; idem, "Are Naturalists Old-Fashioned?" *American Naturalist* 100 (1966): 541–50. Though Dobzhansky was the architect who contributed most to the making of a mechanistic and materialistic science of evoluton, he was also to reject mechanistic materialism most emphatically. Of the unifiers, Dobzhansky was actually the *least* positivistic. In his 1964 article, he held Comte responsible for the hierarchical ordering of knowledge. In limiting mechanistic materialism, Dobzhansky was to closely echo Teilhard de Chardin in his view of a mystical form of humanism. Dobzhansky was a member of the Eastern Orthodox Church. For the most recent discus-

the unifying properties of evolution, Dobzhansky forcefully argued for the importance of evolutionary study to all biologists, in so doing first ushering in the phrase that gained currency with evolutionists in the wake of the modern synthesis:

> I venture another, and perhaps equally reckless, generalization—nothing makes sense in biology except in the light of evolution, *sub specie evolutionis*. If the living world has not arisen from common ancestors by means of an evolutionary process, then the fundamental unity of living things is a hoax and their diversity, a joke.[245]

In 1970 Dobzhansky adapted his earlier *Genetics and the Origin of Species* to incorporate the newer developments in molecular genetics. *Genetics of the Evolutionary Process*, the title of this new book, once again argued for the orderly hierarchy of biology: molecular, cellular, individual, and populational. This hierarchy was responsible for "The Unity of Life and the Diversity of Life," the title of his preface. Closing this, his latest synthetic account of the state of evolutionary biology, Dobzhansky reflected once again on the "Creative Process" of evolution and the origin of "Man." He wrote:

> the living world not only persists but also contains a greater variety of forms and more complex, sophisticated, or 'progressive' ones than in the past. Teilhard de Chardin (1959) has said that evolution is 'pervading everything so as to find everything.' This is an overstatement, since the potentially possible variety of gene patterns is vastly greater than the variety ever realized. Yet he is right in essence: natural selection has tried out an immense number of possibilities and has discovered many wonderful ones. Among which, to date, the most wonderful is man.[246]

Echoing Dobzhansky, "Molecules to Man" and "Unity and Diversity of Life" became the slogans of the biological sciences in the 1960s.[247]

---

sion on Dobzhansky's religious beliefs, see Mark B. Adams, ed., *The Evolution of Theodosius Dobzhansky* (Princeton: Princeton University Press, 1994).

[245] Dobzhansky, "Biology, Molecular and Organismic," p. 449.

[246] Theodosius Dobzhansky, *Genetics of the Evolutionary Process* (New York: Columbia University Press, 1970), p. 431. See the work of his self-recognized "epigone": R. C. Lewontin, *The Genetic Basis for Evolutionary Change* (New York: Columbia University Press, 1974). This book was the twenty-fifth in the Columbia Biological Series and followed the Jesup Lectures that he had given in 1969. Lewontin had followed his mentor's advice on studying variation in populations.

[247] One of the most widely used biological sciences textbooks in the 1960s, produced by the Biological Sciences Curriculum Study (BSCS), was the "Blue" version, which had as its overarching theme "Molecules to Man." The textbook also incorporated the theme of the "Unity of Life and Diversity of Life." See *Biological Science: Molecules to Man*, rev. ed. (Boston: Houghton Mifflin, 1968). See the discussion below.

Evolutionary biology came through this period intact and began to be included in routinized American high school curriculums as it emerged—unquestionably—as the "central organizing principle" of the biological sciences. In the midst of the competitive fears generated by the heightening of the Cold War, American biologists at AIBS, now with their mature science, launched the most ambitious enterprise to reform the educational curriculum of biology in America. With the help of $143,000 from the National Science Foundation (NSF), AIBS formed a Committee on Education and Professional Recruitment in 1955, which in turn organized the Biological Sciences Curriculum Study (BSCS) in 1958.[248] This initiative had followed—without surprise—from earlier reforms in the curriculums of physics and chemistry. At the same time that it would aid the professionalization of incipient biologists and would promote the biological sciences to the growing population of the postwar baby boom generation, it would also sustain belief in a unified science of biology. The centrality of evolution—an official statement of BSCS—was stressed by the BSCS *Biology Teachers' Handbook*: "It is no longer possible to give a complete or even coherent account of living things without the story of evolution," began the official statement; and it concluded with: "evolution, then, forms the warp and woof of modern biology."[249] Designing a popular series of textbooks with accompanying curricular materials that made students experience biology as a laboratory science that stressed "observation and experiment," and at the same time introduce them by the inclusion of historical vignettes to the progressive history and philosophy of the science—replete with "great men of science,"[250] BSCS drew on an impressive number of professional biologists, but especially on the expertise of some the unifiers: G. Ledyard Stebbins, Jr., and G. G. Simpson, as well as Hermann J. Muller.[251] Three colorful textbooks launched

[248] See the first BSCS newsletter for the history and initial statement of the goals of BSCS. Walter Auffenberg, ed., *BSCS Newsletter* 1 (1959): 1–9. For the historical origins, organization, and impact of BSCS, see William V. Mayer, "Biology Education in the United States during the Twentieth Century," *Quarterly Review of Biology* 61 (1986): 481–507; Arnold B. Grobman, *The Changing Classroom: The Role of Biological Sciences Curriculum Study* (Garden City, N.Y.: Doubleday, 1969); Paul DeHart Hurd, *Biological Education in American Secondary Schools 1890–1960* (Washington, D.C.: American Institute of Biological Sciences, 1961). See also Paul DeHart Hurd, "An Exploratory Study of the Impact of BSCS Secondary School Curriculum Materials," *American Biology Teacher* 28 (1976): 79–85.

[249] BSCS, *Biology Teachers' Handbook*, Joseph J. Schwab (supervisor) (New York: Wiley, 1963). As cited in Addison E. Lee, "The BSCS Position on the Teaching of Biology," *BSCS Newsletter* 49 (1972): 6.

[250] BSCS included vignettes from the history and philosophy of science that served as exemplars of the scientific method.

[251] Other biological luminaries who participated include Marston Bates, Daniel Arnon, Garrett Hardin, Joseph Wood Krutch, Alfred Romer, Paul Sears, Philip Handler, Bruce Wallace, Bentley Glass, and John Moore.

in 1960–61 (the "Green," the "Blue," and the "Yellow") were to discipline an entire generation of emerging new professional biologists to the
belief in biology as unified science.[252]

Nor had the efforts to reform and promote the biological sciences been
limited to the United States. In the early 1960s an international push was
made to teach the new biology by the Organization for Economic Co-
operation and Development (OECD).[253] In 1962 some ninety people
from twenty countries who represented "Administrative, University and
School sides of education" met for twelve days near Vevey, Switzerland,
at a seminar on the "Reform of Biology Teaching."[254] A second workshop in 1964 entitled "International Working Session on the Modern
Teaching of Biology" was held at Hellebaek, Denmark. The published
report of the proceedings promoted familiar arguments for the centrality
of biology among the other sciences and urged reforms in its teaching:
"Biology is seen as a unifying element integrating the 'classical' and the

---

[252] In the early 1960s about one hundred thousand employed people could be classed as
professional biologists. For an especially amusing attempt to explain and promote professional biological science in the early 1960s to baby boomers choosing careers, see William
W. Fox, *Careers in the Biological Sciences* (New York: Henry Z. Walck, 1963). He describes
a good candidate in the following terms:

> the candidate must have the curiosity and initiative to work out solutions on his own
> with little guidance after a reasonable period of on-the-job training. The ideal candidate
> for the profession might be visualized as one with a high degree of intelligence, a ten-
> dency to be patient and meticulous, and an intense curiosity about the world around him
> and the beings that live on and in it. His drive and patience must carry him through four
> years of college as a bare minimum, and probably more, for many a ladder of success in
> the biological field has a sign on the top rung that reads 'Reserved for Ph.D.'s.' (p. 18)

The author did not recommend that women (and some men) go into all fields of the
biological sciences: "Technicians who expect to work directly with livestock raisers and gain
their respect should, however, be the 'man's man' type. While female ranchers seem to gain
acceptance, a female range technician probably would find slow acceptance as a consultant"
(p. 76). The pay scales of biologists in the 1960s reflected their relationship to physics and
chemistry. Biologists, psychologists, mathematicians, and sanitary engineers could expect a
lifetime salary of $350,000 to $399,000. Chemists, astronomers, earth scientists, physicists,
and other engineers could expect ot make $400,000 to $449,000. See ibid., p. 98, for full
details. The U.S. Department of Labor *Occupation Outlook Handbook* for 1961 stated that
employment opportunities would be best in biophysics, microbiology, physiology, phar-
macology, and virology (as cited in ibid., p. 101). Fox listed thirty-nine areas that he
defined as "some of the biological sciences." Evolutionary biology was not one. The closest
to evolutionary biology was paleontology and possibly anthropology. Paleontology, systematics, and botany were classed as "thoughtful sciences," "science for the sake of science";
they were of "practical use," but generally were "an incident to the academic value of the
study" (ibid., p. 22). Evolutionary biology never reached the status of "profession."

[253] The OECD was set up in 1960 by member countries of the Organization for European Economic Co-operation and by Canada and the United States.

[254] Papers were given by biologists like T. M. Sonneborn, H. J. Muller, M. Bates,
J. W. L. Beament, and B. Rensch, among others. J. M. Thoday served as president of the
conference. See *New Thinking in School Biology*, report on the OECD Seminar on the Reform of Biology Teaching, 4–14 September 1962.

'scientific' concepts of culture. It should, therefore, be an essential part of the general education of all children and be included in the curriculum of all schools at the upper primary and secondary levels."[255] The report continued:

> Among the sciences, biology has a special claim to the attention of educators, for several reasons. Not only are there obvious considerations (e.g. its educational virtues; man himself being a living organism, the rich fields of experimentation and so on), but also there is the peculiar position which biology occupies in the whole complex of culture. In one direction, it is closely connected with the other natural sciences; in another direction it merges imperceptibly with the social sciences (which might indeed all be considered aspects of human ecology); in other directions it interpenetrates with psychology with philosophy and with ethics.[256]

So successful were these organizational and educational efforts to promote the teaching of the new biological sciences on an international scale that by the late 1960s biology had emerged as the central and most fashionable science. This was especially true in the United States, where the unifying theme of biology was subtended by the unifying theme of evolution. So widespread had the teaching of biology and evolution become that by the 1970s it engendered a reaction from fundamentalists. In the United States—the hotbed of antievolutionary creationism—a series of assaults against the biological sciences and the teaching of evolution began to take place.[257] Moving in, once again, the unifiers addressed the latest threat to their evolutionary science. A BSCS newsletter of 1972 devoted expressly to the teaching of evolution provided the forum for argument with the fundamentalist assaults, and reinforced the official policy toward the teaching of evolution at BSCS. To G. Ledyard Stebbins' reminder that evolution is "the central theme of biology,"[258] and to Bruce Wallace's commentary on "the centrality of evolution to the disci-

[255] "Report of the Working Group," in Paul Duvigneaud, *Biology To-day: Its Role in Education*, report of an OECD Working Session on the Teaching of School Biology, 1964. English version edited by L. C. Comber. Quotation on p. 7.

[256] Ibid., p. 36. The report also added: "We therefore ask not for the addition of more biology to the school curriculum, but for the impregnation of the whole of our education by the subject matter, principles, methods and spirit of biology." Quotation on p. 37.

[257] For a discussion of the creationist resurgence in the wake of the success of BSCS, see George E. Webb, *The Evolution Controversy in America* (Lexington: The University Press of Kentucky, 1994)

[258] G. Ledyard Stebbins, "Evolution as the Central Theme of Biology," *BSCS Newsletter* 49 (1972): 4.

pline of biology"[259] and his claim that "a theory of evolution that is testable by observation and experimentation is imperative if biology is to remain a science,"[260] the chairman of the board of directors of BSCS concluded that in biology "evolution is not only one of the major themes but is, in fact, central among the other themes" and that it had been "organized into a unifying idea widely accepted by modern biologists."[261]

In 1973—two years before his death—Dobzhansky extended the centrality of evolution further still as he defended evolution against the same American fundamentalist assaults, and preserved at the same time the unity of the biological sciences. Writing a lead article featured on the cover of the *American Biology Teacher*, he forcefully restated his earlier assertion that: "Nothing in biology makes sense except in the light of evolution."[262] Adopting this famous phrase, textbooks of biology later in the 1970s and 1980s made the centrality of evolution part of the received and established wisdom of the profession.

The most popular undergraduate textbook devoted expressly to evolutionary biology, Douglas J. Futuyma's *Evolutionary Biology*, stressed the facticity and all-encompassing nature of evolution as the synthesis reigned supreme. Introducing the subject of study in chapter 1, Futuyma wrote: "Evolution in the broadest sense is the observable or inferable fact of change. . . . Thus evolution is all-pervasive. Galaxies, chemical elements, religions, languages, and political systems all evolve."[263] With its "symbiotic relationship" to geology and because "each of the subdisciplines of biology, from biochemistry to ecology, assumes evolution and interprets data in its light," Futuyma stated: "Evolution is, indeed, the one coherent system of principles that unifies all of biology."[264] So strong was this message that the summary ending condensing the point of the first introductory chapter began with a statement reinforcing Dobzhansky's assertion: "Evolution, a fact rather than mere hypothesis, is the central unifying concept of biology. By extension it affects almost all other fields of knowledge and thought and must be considered one of the most influential concepts in Western thought."[265]

At the height of the controversies surrounding evolutionary theory

[259] See the introductory statement to Wallace's article. Bruce Wallace, "Science, Biology and Evolution," *BSCS Newsletter* 49 (1972): 2–3.

[260] Ibid., p. 3.

[261] Addison E. Lee, "The BSCS Position on the Teaching of Biology," p. 6.

[262] Theodosius Dobzhansky, "Nothing in Biology Makes Sense Except in the Light of Evolution," *American Biology Teacher* 35 (1973): 125–29.

[263] Douglas J. Futuyma, *Evolutionary Biology* (Sunderland, Mass.: Sinauer, 1979), p. 7.

[264] Ibid., p. 11.

[265] Ibid., p. 14.

and the modern synthesis of evolution, a second, revised edition of *Evolutionary Biology* made its way to a generation of students eager to learn of a subject that was making scientific and popular news.[266] The new edition not only introduced students to exciting controversies in evolution, but also formally introduced them to the historical event of the modern synthesis, the "founding fathers" (complete with photographs) and their texts, all in a special new section entitled "The Modern Synthesis." The new edition also served to promote the vital evolutionary theory on equal footing with Newtonian theory: "The theory of evolution is a body of interconnected statements about natural selection and the other processes that are thought to cause evolution, just as the atomic theory of chemistry and the Newtonian theory of mechanics are bodies of statements that describe causes of chemical and physical phenomena."[267] The 1974 edition of the "Golden Guide" series transmitted the new evolution to a popular science audience. "Mankind," concluded paleontologist F. H. T. Rhodes,[268] "the product of organic evolution, is now technically equipped with power, if not the will, to control the future development of life on earth. Psychosocial evolution has now displaced the older processes of organic evolution in human communities. Knowledge, traditions, values, and skills are now transmitted from one generation to another through books and teaching institutions rather than being learned anew 'from scratch' by each new individual." With the appearance of the "Golden Guide" to evolution, a popular science audience was to acknowledge the modern synthesis of evolution as complete.[269]

The synthesis itself was logically extended and entered a "new" stage at this time. Emerging from commitments to the late 1950s and 1960s

[266] The early 1980s had seen an explosion of popular science magazines, which served to promote science to vast popular audiences.

[267] Douglas J. Futuyma, *Evolutionary Biology*, 2d ed. (Sunderland, Mass.: Sinauer, 1986), p. 15. The analogy between Darwinian evolutionary theory and Newtonian theory was echoed more recently by some philosophers of biology: "After some general and historical considerations, Darwinian evolutionary theory—a conceptual minefield *par excellence*—is interpreted as a theory of forces not altogether unlike Newtonian mechanics" (Werner Callebaut, *Taking the Naturalistic Turn, or How Real Philosophy Is Done* [Chicago: University of Chicago Press, 1993], p. xvii).

[268] Frank H. T. Rhodes, *Evolution* (New York: Golden Press, 1974). Rhodes had also authored a semipopular book on evolution for the Pelican/Penguin series. See F. H. T. Rhodes, *The Evolution of Life* (Harmondsworth: Penguin, 1962; 2d ed. 1976). Rhodes became the president of Cornell University. It was with his support that the university saw the growth of the history of biology (and the Program in the History and Philosophy of Science and Technology at Cornell). Quotation on p. 153.

[269] Popular science audiences had been responding to the recent developments in evolution from as early as the 1950s. See the corpus of popular work by Loren Eiseley; see also the evolution volume of Time-Life, Life Nature Library series: Ruth Moore, *Evolution* (New York: Time-Life Books, 1962).

adaptationism that upheld the continuum between the gene and the human, there arose a new science from the connection between biology and the human or social sciences: sociobiology. *Sociobiology: The New Synthesis*,[270] the second in a trilogy of new books that appeared in 1975, was written by Edward O. Wilson, one of the next generation of synthesizers.[271] Though this new synthesis could also effectively reduce the human to the object of study of both the biological and social sciences, thus posing the "dilemma" of complete determinism, for Wilson there would be enough "hope" (though "blind hope" still) left behind for human agency and autonomy itself.[272] Other sociobiologists would not agree, and took the continuum to its logical completion: the human was only the packaging for the will of the genes. The genes, now viewed as "selfish," directed the ultimate goal of human beings: to make more copies of themselves![273] Thus human sexuality, community, and all of humanity were effectively reduced to the material world of the genes. Altruistic acts, which could not be accounted for in terms of individual "interests" (in this framework at least), became one of the great problems to be addressed by some of these same sociobiologists; debates over sexual and group selection, rather than individual selection, were to flare up repeatedly following the establishment of the "synthetic theory of evolution" among many of them.[274]

An entire new field of professional historians and philosophers of science (especially of the biological sciences) began to play an important role in unifying the biological sciences. In the mid-1960s, as geneticists

[270] E. O. Wilson, *Sociobiology: The New Synthesis* (Cambridge, Mass.: Harvard University Press, 1975). Wilson defines evolution in mostly genetic terms: "Any gradual change. Organic evolution, often referred to as evolution for short, is any genetic change in organism from generation to generation; or more strictly, a change in gene frequencies within populations from generation to generation" (p. 584). Wilson defines evolutionary biology as "The collective disciplines of biology that treat the evolutionary process and the characteristics of populations of organisms, as well as ecology, behavior, and systematics" (p. 584).

[271] See Wilson's attempts to formulate a synthetic and unifying theory of evolution in his recent autobiographical reflections. Edward O. Wilson, *Naturalist* (Washington, D.C.: Island Press, 1994).

[272] See especially the discussion in E. O. Wilson's *On Human Nature* (Cambridge, Mass.: Harvard University Press, 1978).

[273] Richard Dawkins, *The Selfish Gene* (Oxford: Oxford University Press, 1976); idem, *The Extended Phenotype: The Gene as the Unit of Selection* (San Francisco: W. H. Freeman, 1982). British workers like Dawkins, his contemporary John Maynard Smith, and earlier workers like R. A. Fisher seem especially to lean in the direction of reductionist biological schemes.

[274] For an analysis of these problems, see Mary Bartley, "A Century of Debate: The History of Sexual Selection Theory (1871–1971)" (Ph.D. diss., Cornell University, 1994); and Helena Cronin, *The Ant and the Peacock: Altruism and Sexual Selection from Darwin to Today* (Cambridge: Cambridge University Press, 1991).

like L. C. Dunn and A. H. Sturtevant rewrote the history of genetics,[275] there grew an awareness of the importance of the history of biology. Following also the recent emergence of the history of physics, moves were made to launch the history of biology. With the encouragement of Ernst Mayr, then on the board of syndics of Harvard University Press, Everett Mendelsohn founded a journal expressly devoted to the writing of the history of biology, the *Journal of the History of Biology*. Directly responding to the recent history of physics, the editor's introduction asserts: "While the physical sciences have long served as the paradigm for work in the history of science, and several specialized journals have published articles in this field, this imbalance is now being redressed. Many historians of science are now turning their attention to the complex and often challenging problems of the history of biology, and a new generation of scholars has taken biology as the focus for their historical analyses."[276] The range of articles not only served to counterbalance the recent histories of genetics with consideration of evolution, systematics, and other biological fields, but also provided the vehicle for a new generation of historians and philosophers to consider the biological sciences. Volume 2 of the journal, which included articles from the "Conference on Explanation in Biology: Historical, Philosophical, and Scientific Aspects" held at Asilomar State Park in Monterey, California, in June 1968, begins with an essay setting out the problematic for the subsequent philosophy of biology ("Biology and the Unity of Science") and concludes with the review-article that launched the new philosophy of biology ("What Philosophy of Biology Is Not").[277]

The emergence of a professional group of historians and philosophers who would problematize the unity and diversity of biology emerged just in time, for by the mid-1970s a series of amendments from a group of the younger generation of evolutionists would begin to destabilize evolu-

---

[275] L. C. Dunn closed his short history of genetics with a call to understanding evolution, "a central problem of biology, and in fact biological problems generally," which also "requires cooperation of studies at all levels" (Dunn, *A Short History of Genetics*, p. 220).

[276] Everett Mendelsohn, "Editorial Foreword," *Journal of the History of Biology* 1 (1968): iii.

[277] The conference participants included Mark Adams, Garland Allen, Morton Beckner, Ernst Caspari, Frederick B. Churchill, William Coleman, Ruth Schwartz Cowan, Hubert Dreyfus, Bentley Glass, T. A. Goudge, Clifford Grobstein, Adolf Grunbaum, Keith Gunderson, Thomas Hall, Jonathan Hodge, Frederic Holmes, David Hull, David Kitts, Hugh Lehman, Richard Lewontin, Edward Manier, Ernst Mayr, Everett Mendelsohn, Ernest Nagel, Herbert Odum, John R. Platt, Hilary Putnam, L. J. Rather, Carl Sagan, Kenneth F. Schaffner, Michael Scriven, Dudley Shapere, George Gaylord Simpson, Judith Swazey, and William Weedon. David L. Hull, "What Philosophy of Biology Is Not," *Journal of the History of Biology* 2 (1969): 241–68.

tionary biology—from within the framework the architects had established. Beginning with an initial—and initially somewhat minor—reform of Simpson's paleontological theory, this generation pointed to what they actually "saw" in the fossil record: periods of stasis, followed by abrupt periods of change, rather than smooth, gradual evolutionary change. With their motto-turned-slogan of accounting for evolutionary "stasis" and not just change, this generation unwittingly began to reform—in a profound sense—the synthetic theory of evolution. The "punctuationalist" model of evolutionary change (as it came to be known), extended even further to launch a wholescale "critique of the adaptationist" program, eventually endangered the evolutionary and adaptationist framework that the architects had built. Sundering the continuum between microevolution and macroevolution, and effectively limiting the determinative action of the gene on the human condition, in so doing "breaking" the logical extension to sociobiology, this generation constructed an argument that would begin the process of disunifying the biological sciences—*from within*. Problems of developmental biology that remained (and had never been part of the synthesis) would also be brought to the fore by some of this generation so that Goldschmidt—the former "outsider" to the synthesis—was raised to the status of an antihero and "heretic" in the history of evolutionary biology. So, too, would the "decoupling of macroevolution and microevolution" lend legitimacy to this group's emerging young field of knowledge—complete with its own language (now incorporating the newer "exaptation"), which would be expressed in their new *Paleobiology* journal.[278]

But it was through the selective writing and rewriting of the *history* of biology that the belief in biology as a unified yet autonomous biological science was to be continuously reinforced.[279] Ernst Mayr's own historical and philosophical reflections were to continue to reach wider and wider audiences of biologists in the 1980s.[280] Just as the neutral theory of evo-

[278] The clearest discussion of macroevolution is Steven M. Stanley, *Macroevolution: Pattern and Process* (San Francisco: W. H. Freeman, 1979). The tenth anniversary issue of *Paleobiology* included a special notice of the history of the journal: "PALEOBIOLOGY was founded one decade ago for the publication of research papers and essays in the area of biological paleontology. The purpose of the journal was, and remains, the uniting of paleontology with modern biology" (*Paleobiology* 11 [1985]).

[279] See Kitcher, *Abusing Science*, for an example of how the history of biology and the philosophy of biology sustain the view of biology as unified science. Kitcher writes: "Evolutionary theory is not simply an area of science that has had some success at solving problems. It has unified biology and it has inspired important biological disciplines" (p. 54).

[280] See Ernst Mayr, *The Growth of Biological Thought* (Cambridge, Mass.: Belknap Press of Harvard University Press, 1982); see also his collection of essays (especially "Is Biology

lution had exploded on the scene, punctuated equilibrium was being introduced, and the critique of the adaptationist program was being formulated—all of which began to call for amendments in evolutionary biology that began to disunify biology; just as a totally mechanistic, materialistic, reductionist biology—devoid of progress and purpose and with no meaning to life—was gaining acceptance, Ernst Mayr once again stepped in.[281] Drawing together a group of original participants, scientists, philosophers, and young historians of science, Mayr reintroduced and revivified the belief in a unified science of biology. Borrowing and adapting the title of Huxley's 1942 book, the group concentrated their efforts on figuring out "what actually happened" during the "evolutionary synthesis" between 1920 and 1950. Looking for theories and trying to fix the synthetic theory at the core, they were frustrated by discovering that the synthesis was a "moving target."[282] Spilling gallons of ink on the subject and engaging in heated disputes for nearly a decade, the growing numbers of commentators on what became *the* "synthesis" would only agree in making this count as a historical "event." The Columbia Biological Sciences Series—reissued as "The Columbia Classics in Evolution"— at the same time, with historical prefaces from the next generation of architects-turned-"paleobiologists"—also sustained the belief in biology as unified science. The historical note on the series in the reissue of Dobzhansky's *Genetics and the Origin of Species* begins with the assertion: "Evolution, the proposition that all organisms are related by descent, is the central organizing principle of biology."[283] What the architects of the synthesis had worked to construct had by 1982 become a matter of fact.[284]

---

an Autonomous Science?"), *Toward a New Philosophy of Biology* (Cambridge, Mass.: Belknap Press of Harvard University Press, 1988); idem, "How Biology Differs from the Physical Sciences," chap. in David J. Depew and Bruce H. Weber, eds., *Evolution at a Crossroads* (Cambridge: MIT Press, 1985), pp. 43–63. See his most recent *One Long Argument: Charles Darwin and the Genesis of Modern Evolutionary Thought* (Cambridge, Mass.: Harvard University Press, 1991).

[281] Mayr has repeatedly responded to challenges to the synthetic theory. In addition to works already cited, see his analysis of the Gould and Lewontin critique of the adaptationist program in Ernst Mayr, "How to Carry Out the Adaptationist Program?" *American Naturalist* 121 (1983): 324–34.

[282] See Richard Burian, "Challenges to the Evolutionary Synthesis," *Evolutionary Biology* 23 (1988): 247–69.

[283] Niles Eldredge and Stephen J. Gould, "A Note on the Series," in 1982 reissue of Theodosius Dobzhansky, *Genetics and the Origin of Species* (New York: Columbia University Press, 1982), p. ix.

[284] The centrality of evolution had thus been rendered tacit knowledge, part of the received wisdom of the profession. Here, I am drawing on some of the literature in science studies. See how matters of fact were constructed in the early modern period. Steve Shapin and Simon Schaffer, *Leviathan and the Air-Pump: Hobbes, Boyle, and the Experimental Life* (Princeton: Princeton University Press, 1985).

The long-awaited volume of Mayr's 1974 meeting appeared in 1980 and began to serve as the textbook initiating practitioners into the burgeoning new field of the history of biology.[285] Adopting contextualist historiographic frameworks and examining the interplay of the material, literary, and social practices deeply embedded within the Western mentalité, this generation of architects-turned-historians were to historicize to reconstruct[286] the "meaning of the synthesis." The meaning of the synthesis—to a close reader of texts—has always resided in the textbook title: *The Evolutionary Synthesis: Perspectives on the Unification of Biology*.

[285] Ernst Mayr also produced his monumental *Growth of Biological Thought* at roughly this time.

[286] A construction of a construction, as Clifford Geertz would say.

PART THREE

Persistent Problems

CHAPTER SIX

Reproblematizing the Evolutionary Synthesis

> If we would have new knowledge, we must get us a whole
> world of new questions.
>
> Susanne K. Langer, *Philosophy in a New Key*

AND SO the narrative (more correctly, the postscript) of *Unifying Biology* ends where the author inherited the problem of the synthesis, in Will Provine's 1982 seminar at Cornell University. If it has achieved anything at all, it is the closure of one set of problems encountered there, and the concomitant opening and posing of another set of problems that unfold here.

The construction of a unifying narrative has given some coherence to the evolutionary synthesis, as an interpretive narrative framework has been made available to its students. In this it has ended at least one of the problems that existed in 1982, the absence of any coherent story at all and with it the inability to make any sense of the historical event. But the narrative also comes with a price, as the trajectories or historical threads of the narratives and a range of historical details are pushed aside by the narrative drive of the text. Thus, because the narrative construction of *Unifying Biology* does not permit formal analytic discussion in the text proper, I have added this closing section to answer further questions and add points of clarification for readers. In some cases I offer further questions for exploration. As stated in part 1, *Unifying Biology* should reach a wide audience from the humanities and the sciences. Because these audiences will raise a different set of questions befitting their intellectual predispositions, I respond to each directly. A closing section reproblematizes the project for further studies.

BIOLOGISTS

Many biologists, especially nonevolutionary biologists, will wonder where each of their respective disciplines fits into the evolutionary synthesis. If in fact a proper synthesis took place that led to a unified science of biology by the 1950s, then why were already mature disciplines like microbiology, physiology, and embryology largely absent? Why, for instance, were members from these disciplines not actively engaged in the evolutionary synthesis as a whole? Why, moreover, were some representatives continuing to disparage evolutionary studies until well after the synthesis?

Historians have long noted the absence of embryology in the synthesis, especially because it was through embryology that the first of the "countersynthesis" arguments were launched. Tentative answers have been proposed to understand the role that embryology played in the synthesis. Ernst Mayr has, for instance, noted that embryologists simply did not *want* to be included in the synthesis. The absence of embryology was not simply a matter of being "left out" of intellectual and organizational activities; instead, embryologists consciously sought to avoid serious contact with evolutionists: "The representatives of some biological disciplines, for instance, developmental biology, bitterly resisted the synthesis. They were not left out of the synthesis, as some of them now claim, but they simply did not want to join."[1] This Mayr ascribes to the antievolutionary sentiment among embryologists, whose science was heavily experimentalized and who continued to disparage evolutionary study for its nonrigorous nature. As well, despite their claims to methodological rigor, embryologists were also heavily predisposed to Lamarckian rather than Darwinian explanations (for physiology and microbiology, see the discussion below). Here, the larger framework proposed in *Unifying Biology* can shed some light on understanding the absence of embryology and on Mayr's historical sense that embryologists wanted to be left out for these reasons.

By the mid-1950s, at the peak of the evolutionary synthesis, the architects (including Simpson, Mayr, Huxley, et al.) were overstressing their sense of unity. This was not based on an attempt at self-aggrandizement or self-importance or even to construct an "empty" proevolutionary argument at a time of growing "threat" from molecular biology and biochemistry, but because they were responding to persistent and historically induced charges of disunity. Because the comparative scale of the sciences weighed so heavily in favor of the physical sciences, and because the physical sciences served as the historical, logical, and epistemic exemplars to the biological sciences, biologists rushed to prove to themselves

---

[1] Ernst Mayr, "What Was the Evolutionary Synthesis?" *Trends in Ecology and Evolution* 8 (1993): 31–34. Quotation on p. 32.

(and to others) that a formerly fragmented and immature science had finally achieved the unity to "rival the unity of physics and chemistry" (in Huxley's words). On a comparative scale, the biological sciences could at least *claim* to be unified, a position few could even hope to support in the 1920s; additional support in favor of sociopolitical if not epistemic unity also came with the assembly of the central organizational apparatus of AIBS. As Toby Appel has argued, such a biological organization had not even been possible in the 1920s, the period immediately preceding the evolutionary synthesis.[2]

But as the sense of unity emerged and as biologists sought to organize their community, so, too, did it become apparent—especially to evolutionists at the helm of the unifying movement—that some biological disciplines were "left out" of the synthesis. Thus Mayr's need to respond to embryology with an explanation that embryologists did not want to be included in the larger discussions (the architects actually had little to say to these embryologists). But just because these fields did not appear to be included (or may have even been inimical to evolutionary studies in the 1920s and 1930s) does not mean that they were left out of the synthesis forever or completely. The "absence" of embryology can be understood in the light of *Unifying Biology* as being much more apparent than real. Waddington's 1953 paper historically constructs the first counter-synthesis argument. Historians of science often pick up Waddington's tone of incompleteness and his critique of mathematical population genetics and interpret the entire paper as an antisynthesis argument. Few have noticed, however, that Waddington's notorious article appeared in a volume recording proceedings of a special Oxford conference devoted to the subject of evolution. The volume itself was subsequently entitled "Evolution." This conference had been officially sponsored by the Society for Experimental Biology (the society that Julian Huxley helped found), a society which, according to Mayr, had intellectual interests historically opposing nonexperimental sciences like evolution. Here the effect of the evolutionary synthesis seems to have been lost to historians: prior to the evolutionary synthesis, such a volume, even though it included Waddington's "notorious" critique, would never even have existed; it will be recalled that evolution was not considered a sufficiently experimental or rigorous enough science to be taken seriously by the

---

[2] Toby Appel, "Organizing Biology: The American Society of Naturalists and Its 'Affiliated Societies,' 1883–1923," in Ronald Rainger, Keith R. Benson, and Jane Maienschein, eds., *The American Development of Biology* (Philadelphia: University of Pennsylvania Press, 1988), pp. 87–120.

larger audience of experimental biologists. That the forum for publication exists at all supports the claim that a legitimate science of evolution had emerged during the synthesis that unified enough of the biological sciences to generate discussion of what remained.

The volume also drew on examples from an amazingly diverse range of organisms traditionally not associated with evolutionary study; included here are fungi, plants, and other long-neglected organisms. The volume also included articles by a broad spectrum of biologists not traditionally associated with evolution, including physiologists, embryologists, and biochemists. If these scientists and their disciplines had been completely excluded from the synthesis, as has long been claimed by certain historians and critics of the synthesis, then why was there so much willingness to explore evolution from so many disciplines and drawing on such a diverse set of organisms?

Much of this confusion concerning unification and respective biological disciplines can be cleared up if the drive for unification is seen as a historical *process* (hence also the title of *Unifying Biology*). Once enough unity was attained, discussion turned to what remained to be unified. Haldane's introduction to the above "Evolution" volume is especially interesting, for it points out that the tone of celebration at the Princeton meetings was dampened at the Oxford meetings as workers realized that there was still much to do to achieve a *total* synthesis of biology. Historians could interpret the Oxford volume strictly as a countersynthesis movement, but if they do so they ignore the big historical picture; although there were notes of discord, there were also points of agreement among the workers, or there would have been no conference, no edited volume, and no meaningful dialogue at all.

Waddington's own note of "discord" in 1953 in the context of 1960s developmental biology is understandable.[3] It can best be seen as part of a process of unification that followed after, or in the wake of, the evolutionary synthesis. After the 1950s embryologists converted to "developmental biologists."[4] Weekly/monthly issues of *Science*, *Cell*, or the myriad of molecular biology journals indicate that developmental biologists are working feverishly to unify their field with evolution through the integrative methods of molecular biology. This unificatory drive also attempts to "synthesize" biochemistry and physiology, along with behav-

---

[3] Waddington himself, it will be recalled, wrote his *Evolution of an Evolutionist* to explain his contributions to modern evolutionary biology.

[4] Developmental biology was a category of the life sciences recognized by the National Science Foundation in the 1950s.

ioral ecology and physiological ecology.[5] A recent volume edited by Peter Grant and Henry Horn in honor of John Tyler Bonner (another heir to the evolutionary synthesis), entitled *Molds, Molecules, and Metazoa*, includes a suite of articles that show that an appreciable synthesis among these biological disciplines is continuing.[6] Even more recently a conference held in honor of the fiftieth anniversary of G. G. Simpson's synthesis in *Tempo and Mode in Evolution* discussed the influx of molecular approaches to solve evolutionary questions and whether a new synthesis to achieve Simpson's own was taking place.[7] Judging from just some of this recent literature, arguments for the centrality of evolution, for its unifying properties and its ability to act as a "glue" in the process of unification still continue to hold sway; and new grant programs still draw in broadly trained workers who will continue the integration. An announcement of the "postdoctoral fellowships in molecular studies of evolution," initiated in 1987 and sponsored by Alfred P. Sloan Foundation, justifies the program in the following way:

> Evolution has played a central role in the biological sciences since the time of Darwin. Within the past decade, it has become possible for the first time to unravel the millions of years of evolutionary history encoded within the genomes of living species. The Sloan Foundation believes that although the tools of molecular biology offer exciting possibilities for expanding scientific knowledge about evolution, there are too few scientists trained in the complexities of both evolutionary and molecular biology. This postdoctoral awards program is intended for scientists interested in developing relevant interdisciplinary skills. We particularly wish to encourage postdoctoral molecular biologists moving to laboratories devoted to evolutionary biology, and of evolutionary biologists to laboratories of molecular biology.[8]

[5] For additional attempts to synthesize ecology, physiology, and biochemistry within an evolutionary framework, see P. Calow, ed., *Evolutionary Physiological Ecology* (Cambridge: Cambridge University Press, 1987); Martin E. Feder, Albert Bennett, Warren W. Burggren, and Raymond B. Huey, *New Directions in Ecological Physiology* (Cambridge: Cambridge University Press, 1987); and see the synthesis between evolution and physiology in Brian Keith McNab, *The Physiological Ecology of Vertebrates: A View from Energetics* (Ithaca: Cornell University Press, forthcoming).

[6] Peter R. Grant and Henry S. Horn, eds., *Molds, Molecules, and Metazoa: Growing Points in Evolutionary Biology* (Princeton: Princeton University Press, 1992). The influence of the synthesis on J. T. Bonner is highlighted by the title to his own contribution to the volume: "Evolution and the Rest of Biology."

[7] Jon Cohen, "Will Molecular Data Set the Stage for a Synthesis?" *Science* 263 (1994): 758.

[8] Up to ten fellows were to be selected. The program was designed to last six years. *Announcement of Postdoctoral Fellowships in Molecular Studies of Evolution* (Final Round), held in 1993.

To be sure, what exactly counts as a successful synthesis or a unification is a serious topic for discussion, especially for philosophers of science who continue their debates on the subject of what counts as a proper unification;[9] but judging from biologists' historical meaning of the term and its contemporary use in their journals with respect to development and evolution, it is a possibility that biologists believe can be attained; with respect to the synthesis, enough biologists have traditionally held that it was a proper "synthesis" between genetics and selection theory to continue the process of unifying other regions of the biological sciences. To sum up the discussion on the "absence" of embryology in the synthesis, embryology was not yet ready—for varied reasons—to become part of the unification efforts.

Within the wider picture of *Unifying Biology* one can also situate the science of microbiology within the evolutionary synthesis. Note 204 in chapter 5 of *Unifying Biology* points to the "discovery" of Precambrian microfossils, blue-green bacteria (also known as cyanobacteria and originally termed blue-green algae), and other kinds of cellular life-forms by workers like Stanley Tyler and Elso Barghoorn in the mid-1950s. This was one of the most spectacular discoveries of the twentieth-century biological sciences but has received very little historical attention. Outside the smaller circle of Precambrian specialists, few have considered the historical and philosophical implications of this discovery. Tyler and Barghoorn's work was picked up and extended by cell biologists like Lynn Margulis and then by biochemists like Richard Dickerson and subsequently by others who began to explore evolution at the cellular level, and hence began to think about the cell as the unit of evolution. Having trained in cytology, biochemistry, and microbiology, these individuals reflected also a disciplinary realignment with evolutionary biologists who had previously examined evolution at only the genic, organismic, or populational level; thus the integration that linked gene, to cell, to organism, to populations was taking place in the late 1950s and early 1960s. In this sense these individuals could also be seen to be unifying portions of bacteriology or microbiology with evolutionary biology. The field known as Precambrian paleontology was created as a result. Later it was amended to Precambrian paleobiology with the paleontological reformation of evolution and the invention of "paleobiology" in the 1970s and 1980s.

---

[9] Arguments against the unity of the biological sciences and the unity of the sciences are lively topics for philosophical discussion: see John Dupré, *The Disorder of Things: Metaphysical Foundations of the Disunity of Science* (Cambridge, Mass.: Harvard University Press, 1993); Alexander Rosenberg, *Instrumental Biology or the Disunity of Science* (Chicago: University of Chicago Press, 1994).

Though not all of bacteriology/microbiology can be considered "unified" with evolution, portions of the larger discipline as well as enthusiastic practitioners flourish under the wider rubric of evolutionary biology. Among these heirs to the synthesis are included Elso Barghoorn, J. W. Schopf, and Lynn Margulis (who had trained in biochemistry/cytology). Other heirs to the synthesis, who linked their interests to evolution as part of the unifying process, are Melvin Calvin, Sidney Fox, Cyril Ponnamperuma, Joshua Lederberg, and Carl Sagan. Each of these individuals and their respective disciplines was part of the unifying process during the late 1950s and 1960s and consider their work as informing some critical portion of the evolutionary narrative. Although they and their disciplines were not initially part of the synthesis of the 1940s, they were increasingly active in evolutionary circles in the 1960s (Carl Sagan's first published scientific article, it will be recalled, was in *Evolution,* the journal that Ernst Mayr helped found). In like manner the synthesis of molecular biology and evolutionary biology as told in the narrative of *Unifying Biology* incorporated the molecule within this scheme, though this took place through largely heroic efforts on the part of workers like Richard C. Lewontin (yet another heir to the evolutionary synthesis) as well as the promotional efforts of Dobzhansky, Simpson, and the other architects.

The recent recognition of the rapidity and extent of microbial evolution leading to antibiotic resistance, as well as to the evolution of new pathogenic organisms, demonstrated alarmingly by "emerging viruses" like *Ebola zaire*, has led to one of the loudest calls in recent scientific history to unify the principles of evolutionary biology with those of the health sciences. The new area of knowledge thus resulting in "evolutionary epidemiology" or "evolutionary medicine" would also "help raise evolutionary biology's impoverished status in the minds of many outside the field" (not to mention the possibility of receiving no less a recognition than the Nobel Prize for successfully wedding the two).[10] There is little question that this will prove one of the most fruitful and important areas of integration and exchange. Undergirding the logic of its existence

[10] See the introduction to Paul W. Ewald, *Evolution of Infectious Disease* (Oxford: Oxford University Press, 1994), p. vi. See also Randolph Ness and George C. Williams, *Why We Get Sick: The New Science of Darwinian Medicine* (New York: Times Books/Random House, 1994); Stephen S. Morse, ed., *Emerging Viruses* (Oxford: Oxford University Press, 1993); idem., *The Evolutionary Biology of Viruses* (New York: Raven, 1994); Laurie Garrett, *The Coming Plague* (New York: Farrar, Straus, Giroux, 1994). See also Marc Lappé, *Evolutionary Medicine: Rethinking the Origins of Disease* (San Francisco: Sierra Club, 1994). Books on evolution are also including the growing synthesis between evolutionary theory and disease: see chap. 18, entitled "The Resistance Movement," in Jonathan Weiner, *The Beak of the Finch: A Story of Evolution in Our Time* (New York: Alfred A. Knopf, 1994).

is the evolutionary synthesis, the belief in the unification of biology, science, and all knowledge.

The case of anthropology's relation to the evolutionary synthesis, or notably its absence, is worthy of discussion. Given anthropology's present interests in "human evolution" one would have thought that anthropologists had been active members of the evolutionary synthesis, yet few were involved even informally, let alone actively. Although the connections for the linkage between anthropology and genetics had been fortified by Dobzhansky, it had not drawn in workers until the mid-1950s, when largely through the organizational efforts of some anthropologists like Sol Tax, anthropologists began to include biological anthropology as part of their professional interests. The historical conditions under which anthropology entered the evolutionary synthesis and how the reconfigurations leading to biological anthropology and then human biology took place in the 1960s and 1970s is clearly a significant project for future historical inquiry.[11]

How the relationship between evolutionary biology and its "disciplinary other," ecology, developed and how the present coupling took place historically, institutionally, and intellectually is yet another noteworthy subject of discussion. Even more immediately noteworthy is the absence of ecology (and ecologists) in the evolutionary synthesis. Given this present-day coupling, and given that Darwin had integrated "ecological" with his evolutionary thinking, this is a surprising turn of events. It has not, however, escaped the focused attention of some historians of biology, who have spent considerable energy on this question. An entire issue of the *Journal of the History of Biology* was devoted to exploration of the relations between ecology and evolutionary biology (E&EB), the role that ecology played in the synthesis, and to the synthesis between ecology and evolution (leading to evolutionary ecology in the late 1950s and early 1960s).[12] According to the contributors, the absence of ecology was due to reasons ranging from the struggle that ecology faced in becoming an autonomous scientific discipline and in so doing distancing itself from

[11] Sol Tax's organizational efforts to bring anthropology into the evolutionary synthesis are discussed in Vassiliki Betty Smocovitis, "Celebrating Darwin": paper delivered at the department of history and sociology of science at the University of Pennsylvania, and the Boston Colloquium for the History of Science on "The Construction of Scientific Memory" (forthcoming).

[12] See the special issue "Reflections on Ecology and Evolution," *Journal of the History of Biology* 19 (1986). Contributions were by Jane Maienschein, James P. Collins, John Beatty, William Coleman, Joel B. Hagen, William C. Kimler, Sharon E. Kingsland, Richard E. Michod, and Douglas J. Futuyma (who served as commentator). The papers grew out of a conference held in honor of Arizona State University's centennial celebration in 1984.

areas like systematics; to the unpopularity of natural selection and adaptive evolution before the synthesis, which hindered the emergence of evolutionary ecology; to the fundamental philosophical differences in ecological areas of inquiry (many ecologists are concerned with proximate or functional problems while evolutionary ecologists must also concern themselves with ultimate or historical problems).[13] Despite this apparent separation in the 1930–50 period, ecology and evolution did unify—and with much fanfare—by the mid-1960s.[14] But at the same time that ecology and evolution were undergoing their synthesis, ecosystems ecologists, more concerned with systems and understanding proximate causation, grew farther apart. As Joel Hagen points out in his history of ecosystems ecology, in a chapter entitled (appropriately enough) "Evolutionary Heresies," the much desired dream of a grand unified theory of ecology (part of the "New Ecology") has largely failed to be achieved. This has led to the creation of two distinct areas of ecological research: evolutionary ecology (concerned with populations of organisms in historic time) and systems ecology (concerned with functional rather than historical aspects of ecosystems).[15]

To be sure, not all members or components of the disciplines represented by all biologists were "unified" or were completely integrated within evolutionary biology; but enough of a sense of unity among these evolutionary communities had arisen to keep the hope and promise of future unity and dialogue going (despite frequent points of controversy). On a related note, not all the disciplines or members long held to be integrated, including botany, paleontology, and systematics, recognized the evolutionary synthesis; not all botanists, paleontologists, or systematists jumped on the evolutionary bandwagon driven by Stebbins, Simpson, and Mayr. While the proportion of individuals who upheld the evolutionary synthesis in each of these biological disciplines compared to those who rejected, dismissed, or ignored the evolutionary synthesis is

[13] By far the two best assessments of ecology and the evolutionary synthesis are James P. Collins, "*Evolutionary Ecology* and the Use of Natural Selection in Ecological Theory," *Journal of the History of Biology* 19 (1986): 257–88; and William C. Kimler, "Advantage, Adaptiveness, and Evolutionary Ecology," *Journal of the History of Biology* 19 (1986): 215–33. See also the superb commentary on the suite of papers to the entire volume: Douglas J. Futuyma, "Reflections on Reflections: Ecology and Evolutionary Biology," *Journal of the History of Biology* 19 (1986): 303–12. See also Marston Bates's reflections at the Darwin Centennial Celebration: Marston Bates, "Ecology and Evolution," in Sol Tax, ed., *Evolution after Darwin* (Chicago: University of Chicago Press, 1960), vol. 1, pp. 547–68.

[14] See n. 13 for literature detailing the history of evolutionary ecology.

[15] Joel B. Hagen, *An Entangled Bank: The Origins of Ecosystem Ecology* (New Brunswick, N.J.: Rutgers University Press, 1992). Hagen explores how ecosystems ecology and evolutionary ecology "parted company" in his chap. 8, entitled "Evolutionary Heresies."

not known, and requires further examination, it would not be surprising if in fact more were unwilling to join or belong to the evolutionary bandwagon than those who quickly embraced evolutionary perspectives on their fields.[16]

Another point that needs mention here is the fact that many of the disciplines "left out" (with the exclusion of anthropology) may be viewed as fitting into Mayr's philosophical category of those sciences that deal with proximate rather than ultimate causes (hence the grouping embryology, physiology, and microbiology and others). Assuming that Mayr's categories have validity within the biological sciences (this is a subject worthy of further discussion in itself), it should come as little surprise that it is those disciplines (or sciences) that most closely dealt with ultimate causes that were readily "brought into" synthesis. Thus, that a historical cosmological narrative dealing with the origin story was a result of the synthesis is hardly surprising, for it is by definition such historical narratives that explain ultimate causation that concern evolutionary study.[17]

To return to the original issue of the absence or lack of certain biological disciplines in the synthesis: it is important to note that the process of unification is not only part of a historical process but is also actively continuing. Depending on the point of view of the analyst/observer, arguments for and against inclusion/exclusion of disciplines and members can effectively still be made. Taking such a position can also address the other question that biologists reading *Unifying Biology* may have: "To what extent *is* biology a unified science?" Philosophers might answer this question with a simple affirmative or negative and then muster evidential support for either position. But evolutionary biologists and intellectual historians more accustomed to historical (and historicist) thinking might note that *Unifying Biology* stresses the processual features of unification: in this view, final unification can never be attained, for it is not an end but a process.

HISTORICAL ACTORS AND INTELLECTUAL HISTORIANS OF SCIENCE

Another perplexing feature of *Unifying Biology* is the introduction of what I term the positivist theory of knowledge as a backdrop to the

[16] This was certainly true for botanists; very possibly this was also true for many paleontologists.

[17] Ernst Mayr has written extensively on the ultimate–proximate distinction in the biological sciences. The actual mingling of ultimate and proximate causes in the disciplines "brought" to synthesis deserves closer examination.

evolutionary synthesis. Questions may also arise about the often-used phrase "Western Enlightenment thought" and why it is included for an event that took place between the 1920s and the 1950s. Concern with these two features should come from intellectual historians within the history of ideas tradition (this includes also "internalist" historians of science), as well as two of the architects or historical actors. Both G. Ledyard Stebbins, Jr. and Ernst Mayr have read early drafts of *Unifying Biology* and have commented to varying extents on its contents. Mayr has responded formally to an early version of the narrative that appeared in the *Journal of the History of Biology* in a review-essay entitled "What Was the Evolutionary Synthesis?" that appeared in *Trends in Ecology and Evolution*,[18] while Stebbins communicated his comments in personal correspondence.

The introduction of positivist philosophy raises an especially interesting set of questions, because scholars of the synthesis have understood the scientific philosophies of the architects, if they have explored their philosophical backgrounds at all, as being so exceedingly complex and individualistic that they defy any general attempt to place them within a unified philosophical packaging; no great pattern has appeared to exist at all. With respect to contemporary philosophical movements associated with the school of logical positivists, moreover, the architects, if they even addressed this literature (or the representatives) at all, leaned toward an antipositivistic stance, for clearly the biological sciences and especially evolution appeared to defy simple physicalist or theoretical reductions. Consistent with this is the fact that there were few biologists within the "core" group of the Vienna Circle.[19]

Complicating this philosophical perspective of the architects further is the fact that an entire generation of philosophers of biology has formulated its identity in opposition to the extreme physicalist and positivist position. For philosophers of biology, *any* implied or actual connection between the logical positivists of the Vienna Circle and its offshoot, the Unity of Science Movement, and twentieth-century evolutionists seems a very strange conjunction. Why, then, include positivist philosophy in this story at all? If, moreover, what I am claiming is that positivist ideas "originated" from the logical positivist philosophers within the Unity of Science Movement and then directly "influenced" the architects to unify the biological science, then where is the evidentiary support for such a

---

[18] See n. 1.

[19] The absence of biologists in the core group of the Vienna Circle has long been noted by Gerald Holton and others. The spin-off of the Unity of Science Movement did, however, contain biologists like H. S. Jennings, J. H. Woodger, and Sewall Wright. Haldane contributed a paper to one of the congresses, it will be recalled.

claim? Quite rightly so, empirically minded intellectual historians and internalist historians of science will demand an evidentiary base to support such connections between what appear to be unrelated groups and their philosophical or scientific projects. Such evidence would also have to be especially convincing because at least one historical actor, Ernst Mayr, has explicitly stated in response to *Unifying Biology* that the unification of biology "was not an objective in the minds of any of the architects of the synthesis during the 1930–1940 period." Instead, "They were busy straightening out their own differences and refuting the anti-darwinians to have time for such a far-reaching objective. It wasn't until the 1950s when most of the previous difficulties had been resolved that one could begin to think seriously about the role of evolutionary biology in the whole of biology and about the capacity of evolutionary biology to achieve a unification of the previously badly splintered biology."[20]

For good reasons Mayr has also pointed out that evolutionists were too busy straightening out their differences before they could even think about far-ranging objectives like the unification of science or knowledge. But here Mayr's message reveals an especially interesting body of assumptions: though each historical actor may have been unconscious or unaware (to phrase it) of the immediate movements toward unification in relation to his own interests, each held to a tacit belief in the unity of knowledge as a whole. Why else would they believe that they *could* "straighten out their differences"? Mayr's reflection here is especially revelatory, for it demonstrates effectively how tacit the belief in the unity of knowledge, the search for a common ground, and the belief that one could hold a shared universal language—with enough effort—was a possibility for all workers interested in evolution; this meant that all evolutionists could and should come to consensus because all had made an agreement (fundamental to their scientific project) that there was one straight story to be told. (Multicultural theories of knowledge with their plurality of stories would thus pose a problem for such a scientific project.)

Here it is important to note the varying and various streams of positivism that gained currency, come to play, or surface in portions of the narrative of *Unifying Biology*. But rather than exploring definitional components of various schools of positivism with each of the historical actors, which would require individual analytic essays, the grand narrative here uses positivism as the background legitimator to the scientific proj-

---

[20] Mayr, "What Was the Evolutionary Synthesis?"

ect. These "positivistic currents" or threads enter the narrative, and appear to follow roughly the chronology of the positivist genealogy (Comte, Mach to Vienna Circle); but by far the dominant philosophical framework against which biology and evolution are legitimated most closely resembles the initial nineteenth-century positivism that had been assimilated in textbooks of science by the early years of the twentieth century. For similar reasons, the Enlightenment project is introduced in the narrative. Belief in a rational scientific method, progress, liberalism, humanism, and other such elements usually linked up with "the Enlightenment" undergirds the project of *Unifying Biology* along with positivism.

As *Unifying Biology* points out, furthermore, many of these cojoined beliefs (liberalism, science, progress, etc.) run even deeper within many of the canonical texts or great works of the West. The beginnings or origins of Western scientific thought in this narrative traditionally are traced to ancient Greek or Pre-Socratic philosophy (belief in the unity of knowledge, as noted in *Unifying Biology*, is included in Plato's works). Because the grand narrative of *Unifying Biology* begins with Western science itself (more accurately, as retold by historians in the wake of the Enlightenment), the evolutionary synthesis is viewed as the fulfillment of a project that begins with the origin of science and the origin of the "West" itself. Certainly textbook writers of evolutionary biology like Herbert J. Ross made such claims as early as the 1960s. But instead of beginning with this "originary" point in the grand narrative (which is an even more hopelessly ambitious project), *Unifying Biology* focuses on the interval of time between 1920 and 1950, and through the abbreviated postscript extends the narrative to the present so that readers can see the genealogical continuities or discontinuities in the filiations of problems, beliefs, and practices in evolutionary science (also to end the story where it began and situate the author).

Within this interval of time, *Unifying Biology* reflects on the transmission of deep or tacit beliefs, especially through textbooks, schools of instructions, intellectual genealogy, and other sources of acculturation that codify the beliefs of scientists or practitioners, but does not delve too deeply into this subject because it would interfere with the narrative construction of the text. Thus, textbooks of biology have been examined (in background research) at stages from 1900 to the present, but I do not devote prolonged discussion to them.

It should be noted that the "connection" between positivist philosophical movements and the architects is complex and reflects a multidirectional traffic of influences. One can draw some such "connections" between the mathematical modelers (Wright, Fisher, and Haldane) and an intellectual gadfly like Julian Huxley and some of the positivist movements. For instance, as noted in *Unifying Biology*, J. B. S. Haldane participated at least once in the Congress for the Unity of Science. Huxley carried on an active correspondence with Russell; and both Haldane and Huxley were aware of the "Biotheoretical Gathering" that included British theoretically inclined biologists like Woodger and Needham. Even closer to home, Sewall Wright had routinely met with a group of participants in a unity of science or philosophical seminar organized by Charles Morris at the University of Chicago in the mid-1930s. One could draw such "causal" influences among all these individuals, especially when one takes into account how small the community of scientists was in the interval of time between 1920 and 1930, but this would be unnecessary for the immediate project of *Unifying Biology* (though this project is worthy of further exploration), for the goal here is to tell a meaningful story at the disciplinary level. Thus, because it focuses on the collective enterprise (as a whole), the specific positions of each of its individual members need not reflect the trajectory of the discipline, which moves toward greater unity as part of a larger movement toward the unity of science and knowledge, and possibly even sociopolitical and global unity (this certainly was Julian Huxley's much desired goal).

On a related note, some readers will wonder to what extent attending a conference or orchestrating gatherings or societies to bring workers together in person would affect the science involved or the development of participants' points of view. Here the best response came from Charles Morris in response to a critical query he received on the validity of gatherings and movements such as the Unity of Science Movement for unifying science: "I do not believe that the Unity of Science Congresses can be the main source for the unification of the sciences. I quite agree that the basic drive comes from the work of scientists who deal with problems which carry them over the traditional boundaries between the sciences. And I also agree that most of the people who talk about science have not made concrete contributions to the actual unification—though many of the men who have participated in our Congresses certainly are not in that class."[21] I would also note that such groupings and regroupings, even though they may appear to raise controversies, also serve as occasions for the construction of a disciplinary discourse, shared by the community of

---

[21] Letter from Charles Morris to M. King Hubbert, 31 May 1941, Box 2, Folder 1, "The Unity of Science Movement Papers," Joseph Regenstein Library, University of Chicago.

workers (students in classes have been known to "bond" through repeated exposure and physical proximity to each other as they create a culture of their own).

Though the Unity of Science Movement swept through intellectual circles and its products were widely discussed and reviewed, there is little direct evidence that the second grouping of architects like Dobzhansky, Mayr, Stebbins, and Simpson was immediately aware of its philosophical aims and actively working to unify their science of biology as a discipline and at the same time unify it with other sciences. Well after the society was founded, and only after enough of a sense of unity arose, did some of the architects begin to explore the philosophical underpinnings of evolution and biology, especially as both started to come under scrutiny by philosophers of science like Marjorie Grene, Karl Popper, Felix Mainx, Morton Beckner, and Ernest Nagel, all of whom began to take center stage in philosophical circles. Here Mayr's reaction to the philosophy of science based on physics is representative (possibly even constitutive) of the disciplinary reaction; though he was interested and supportive of the Unity of Science Movement, read members' works, and followed their own evolution in the 1950s, he explicitly rejected their extreme physicalist project, which would have effectively reduced biology to physics and also simply did not apply to the biological sciences. Mayr's philosophy of biology (so to some extent Dobzhansky and Simpson, who wrote philosophical treatises on their science) reacted against the strict physicalism as embodied in the works of Ernest Nagel. Specific responses to Nagel's work in the early 1960s, the philosophical criticisms of evolutionary theory, so well foregrounded by Marjorie Grene's early work in the philosophy of science, and how this work fueled the philosophy of biology deserve fuller explication; they should ideally be the subject of yet another work. For our disciplinary purposes, however, it should be sufficient to note that following the grand narrative of the history and philosophy of science (as it had emerged in the post-Enlightenment histories and philosophies of Auguste Comte and William Whewell),[22] biology had to be logically grounded in and reducible to physics (and chemistry), but also had to have some measure of autonomy in order to preserve its independent status.

The introduction of the "positivist" theory of knowledge in *Unifying*

---

[22] Differences between the two and their appeal to different audiences are discussed by Rachel Laudan, "Histories of the Sciences and Their Uses: A Review to 1913," *History of Science* 31 (1993): 1–34. Laudan notes that George Sarton was a positivist while Alexander Koyré was an idealist.

Biology grew out of the contextualist scaffolding that supported the narrative because the historiographic aim was to preserve enough of the perspective (or "voices") of the historical actors. Readers here should return to the discussion in part 3 to reread the hypothetical story that could have been told about the formation of the discipline of evolutionary biology: the group could have come together to form an organization based simply on the "interests" that evolutionists had at a time that they felt themselves threatened from biochemists, geneticists, and later molecular biologists. The answer to the question of what pulled the group together could have thus simply been that they were safeguarding their interests. This scenario would have supported an "externally driven social interest" model for science (the alternative was that the group coalesced simply because their science "came together" when they removed their barriers to synthesis). But in the narrative offered in *Unifying Biology* the response to the question of what pulled the group together is answered by the narrative that was driving the group to performance—a script that had emerged from Enlightenment thought that ran itself through the historical actors. The context of legitimation was thus not solely based on social interests, but on a deeper, narrative-constituted, epistemic context. In support of this argument it will be recalled that evolutionary biology never reached the status of administrative department or category of federal research or even a "profession," yet it did clearly earn itself legitimacy as a proper science (and a highly visible one at that).

To sum up, the response to the question "Why does *Unifying Biology* introduce the positivist theory of knowledge?" is that it views science as a narrative-constituted (or discursively based) epistemic framework that is moving the historical actors as scientists to performance; in other words, science is story or narrative-constituted practice.

Remaining Questions

If science is narrative constituted, philosophers of science will want to know how one can discriminate between stories or whether all stories will hold true (another way of phrasing the problem of relativism). The response here is to state that while science may be narrative-based activity, this does not necessarily mean that all narratives will do. The key question is how narratives are reworked within sets of validating or evidentiary constraints. In other words, the question is how the interplay between "material evidence" and "theoretical explanation" takes place; this question is especially critical for historical sciences, which appear more obviously narrative-constituted (e.g., sciences like evolutionary biology, archaeology, cosmology, and historical geology). Interested readers may note that one philosopher has turned her attention to this issue explicitly and is examining the interplay of such narrative frameworks and the material evidence constituted in anthropology.[23] Another is explicitly exploring the interplay of discursive and nondiscursive components of science to explore meanings of practice.[24]

Another criticism raised by scholars of the synthesis and reinforced by Mayr in 1993 in his response was the emphasis placed by analysts on the synthesis as an Anglo-American phenomenon. As Mayr has aptly stated, *Unifying Biology* does stress Anglophone contributors, possibly to the exclusion of other national contexts of evolutionary activity. Here I wish to respond with an affirmative answer to the well-worn question of whether the synthesis was an Anglo-American historical event; I would also add that such an assertion in no way excludes the contributions of other national contexts of activity. One reason for the emphasis on the Anglo-American context is that it is in Britain and the United States (increasingly the latter) where the institutional and administrative structures for the science or discipline of evolutionary biology were being actively assembled. As the United States became a center for scientific research immediately following the Second World War, it began to draw on its new immigrant scientists to ground its scientific base of operations as well as channeling significant resources into science, a practice that had potential to work in the nation's best interests. In addition to being the location for the founding of the Society for the Study of Evolution, the United States was also the home base for the international journal *Evolution* and the site of the Darwin Centennial Celebration at the University of Chicago in 1959. Interestingly enough, American organizers had a dispute with their British counterparts over who would hold the official Darwin celebration. Following negotiations, the University of Chicago invited Huxley to become a visiting professor at the same time that it was sponsoring its Centennial Celebration. They further legitimated their celebration program by inviting the grandson of Charles Darwin, Sir Charles Darwin, as honorary speaker. If that were not enough of a British carryover to the American Midwest, the conference organizers had considered inviting Winston Churchill (but decided against it because he would upstage the celebrations); instead, they invited H. R. H. Prince Philip

[23] Alison Wylie, *No Return to Innocence: Philosophical Turnings in American Archaeology* (Princeton: Princeton University Press, forthcoming).

[24] Joseph Rouse, *Engaging Science* (Ithaca: Cornell University Press, 1996).

(who declined because of previous commitments).[25] In short, by the late 1950s and early 1960s the United States had already emerged as an international center for evolutionary and biological study; though there were evolutionists all over the world actively enrolled in the new discipline, the site of action shifted to the United States during the evolutionary synthesis.

As he has pointed out, especially in his more recent essays,[26] Ernst Mayr has been right to question his own former understanding of the synthesis as unifying biology successfully. In addition to the challenges posed by the remaining disciplines of knowledge, Mayr has pointed out that the architects of the synthesis failed to agree on one critical point: the target of selection. Whereas naturalist-systematists believed the target of evolution was the individual, geneticists focused on the gene. While the former were interested also in explaining mechanisms for generating diversity in time and space, the latter group took less of an interest in the historical process of evolution. This disagreement had been watered down in the 1940s as the architects believed themselves to be unified with respect to the target of evolution, but has been a persistent problem that has led to some major controversies in evolutionary theory, including the espousal of molecular evolutionists that neutral evolution occurs at the molecular level. Population-oriented naturalists looking to the individual within populations quite rightly resist accepting neutral theory easily, as they both still refer to quite different targets of evolution.

Interestingly enough, in light of some of these persistent controversies and the fact that unification, in the sense of the logical positivists, has led to arguments that biology is reducible to physics and chemistry (at least to highly visible questions of such claims),[27] Mayr has returned to the evolutionary synthesis with some questions about the nature of the unification. In one such response, he explicitly noted: "Historians (perhaps even Mayr and Provine) have overemphasized the unity achieved by the synthesis." [28] This is a far different cry from the position Mayr advocated in the mid-1970s, as he called together participants, historians, and others to turn their mental energies to the subject of the synthesis in

[25] The Darwin Centennial Celebration is discussed in Smocovitis, "Celebrating Darwin."

[26] See n. 1; and Ernst Mayr, "The Modern Evolutionary Theory," paper delivered to the American Society of Mammalogists, June 1995.

[27] John Maddox, "News and Views: Is Biology Now Part of Physics?" *Nature* 306 (1983): 311. The subtitle said the following: "Reductionism is almost a dirty word, especially in biology, but after thirty years of DNA, it is high time that biologists paid attention to the question of what constitutes an explanation."

[28] Mayr, "What Was the Evolutionary Synthesis?" p. 32.

order to conclude that it represented "perspectives on the unification of biology." Thus, Mayr is a living testament that evolutionists are, like all historians, constantly rewriting the past in light of the present.

REPROBLEMATIZING THE SYNTHESIS: CLOSING THOUGHTS

The above sections have discussed some of the critical problems that may concern readers or future historians of the synthesis. Problems remaining to be addressed are the specific fates of certain disciplines and the extent to which practitioners felt a part of their disciplinary culture (or how much they ascribed to the process of unifying their discipline of knowledge). Although they may not necessarily affect our understanding of the trajectory of the discipline of evolutionary biology as a whole, the individual worldviews of each of the architects also deserves closer attention. There is some sign that leading thinkers like Huxley and Dobzhansky, whose existential concerns were entwined with their scientific beliefs in evolution, are now being viewed in wider sociopolitical and philosophical contexts.[29] Picking up on this latter point, the history of philosophical movements like "positivism" and the traffic of influence between the varied biologists and life scientists deserves further study. So, too, postsynthesis developments in areas like sociobiology and reactions/responses to neutral theory and other more recent controversies are all worthy projects for further study.

Methodologically, an examination of different schools of cultural/critical/literary theories and their application to the enterprise we call "science" also needs further exploration. The reconfiguration and restructuring of disciplines—or, alternatively, the lack of such a dynamic in disciplinary structure—also need noting and examination. Similarly, the extent to which we may view science as a culture and as discursive activity also needs further consideration. Along these lines, how the interplay of discursive and nondiscursive practices of science takes place, along with the recognition that "practice" is not an unproblematic term in the cultural study of science should also be important in further discussions.[30] How "communication" occurs, and whether it can occur between,

[29] C. Kenneth Waters and Albert Van Helden, *Julian Huxley: Biologist and Statesman of Science* (Houston: Rice University Press, 1992); and Mark B. Adams, *The Evolution of Theodosius Dobzhansky* (Princeton: Princeton University Press, 1994).

[30] See the excellent discussion in Rouse, "The Significance of Scientific Practices," chap. in *Engaging Science*.

within, and across the disciplines of knowledge, is another key problem
to be addressed.

This project owes a great deal to these more methodological or theoretical areas of inquiry, and it is here that I would like to end; as I reach
completion of this section, I also realize the incompleteness of the project
as a whole and the enormity of the problems that it raises. But I can at
least take comfort in some of playwright Bertolt Brecht's thoughts in his
poem entitled "About the Way to Construct Enduring Works": those
projects that are "devised for completeness" are the ones that "show
gaps," and those projects "planned on a really big scale" are the ones that
are "unfinished."[31]

[31] I thank Gary Weisel and the students in HIS 6486 ("Seminar in the History of Modern Biological Sciences") for the gift of this poem.

CHAPTER SEVEN

Epilogue

> Yet ideas of unity amid diversity, of order amid change, have
> also long been growing, even finding expression, and this not
> merely, as sporadically in all ages, in impressions and
> speculations on decline or on better things; but in clearer and
> more comprehensive surveys of the processes of change, even
> inquiries into its method. These, in fact, have gone towards
> making up that general idea we now more or less share, of
> the universe as not only orderly, but in the process of
> change. Changing order, orderly change, and this
> everywhere—in nature inorganic and organic, in individual
> and in social life—for this vast conception, now everywhere
> diffusing, often expressed, rarely as yet applied, we need
> some general term—and this is Evolution.
>
> —Patrick Geddes and J. Arthur Thomson, *Evolution* (1911)

I HAD chosen the quotation by Geddes and Thomson to end the presentation of *Unifying Biology* because it seemed to echo so many of the polarizing themes present in the story of evolution: unity and diversity, order and disorder, organic and inorganic, individual and society. Its choice of metaphor and of specific language, furthermore, seemed to represent so much of what I wanted to convey in my own view of scientific knowledge: growth, expression, process, universe, and evolution itself. It also seemed to capture the heroic and poetic spirit that has lain submerged in the study of evolution. With the exception of the life writings of Stephen Jay Gould, Edward O. Wilson, and Douglas J. Futuyma (all creatures of grand narrative), the all-encompassing grandeur of the study of evolution, so well summarized in the above epigraph, has been nearly forgotten in the stampede to molecularize its study and to strip it of its metaphysical foundations. Philosophers and historians, who otherwise could uncover and then recover its metaphysical foundations, long ago disengaged science from these metaphysical moorings.

I had also thought that I could end with an optimistic note in much the same way that Darwin had ended his rather unsettling theory with his beautiful and moving "entangled bank" paragraph. But if I am to be honest to the next generation of historians of science, biologists, and other scholars of the evolutionary synthesis reading this text, I cannot conclude with much optimism and without considerable misgivings. Epistemically astute readers will have already guessed reasons for my unease with some issues raised by the process of untangling the synthesis and writing *Unifying Biology*, and what lays in store for the political, ethical, and existential worlds of its readers.

Though it may appear far removed from the internal concerns of evolutionists, the political situation in too many parts of our world bring the multicultural theory used in *Unifying Biology* down from the heavenly realms of academic discourse into the hellfire of sociopolitical reality. At the time of writing, what used to be the "Eastern" bloc of European nations, formerly under the totalizing unity of the Soviet Union, has given way to unbridled movements for ethnic exclusion/inclusion that are among the most violent of the twentieth century (a century now recognized as unsurpassed for its violence and carnage).[1] On the continent of North America, the fragile unity of one of the first of the multicultural nation-states, Canada, threatens to shatter as ethnic groups use arguments for "special status" to raise themselves above their Canadian others. On the silver screens and cathode-ray tubes that mirror ourselves, the images relayed by social commentators like Spike Lee (as in his film *Do the Right Thing*) have demonstrated powerfully, and painfully, what the multiculturalist movement toward diversity looks like at close range: hardly the idealized, harmonious "cultural mosaic" that one hears so much about. Closer to home, in academic departments at American (and other) universities, colleagues who used to strive to find common ground, to build consensus, and to collaborate with the goal of transcending individual differences, now argue in the most grim terms in favor of their individual "interests" as they seek to subvert agreement, unity, and the very existence of a common ground. No advocate or fan of the "white male" and other forms of control and subordination, no real fan of totalizing systems of thought or grand theories, and no fan of the societal compromises that exclude personal freedoms, I am equally saddened to contemplate what appears to be the alternative we have found and set in motion: another equally unlivable worldview. If this is what the narrative of the West, now fragmented in its postmodern, poststructural, post-Enlightenment, and postpositivist condition offers, then the

---

[1] Eric Hobsbawm, *The Age of Extremes: A History of the World, 1914–1991* (New York: Random House, 1994).

next millennium holds little promise of harmonious coexistence, communication, and peacefulness; small wonder that inhabitants of the late
twentieth century experience alienation, a loss of community identity,
and a feeling that societal and personal structures have collapsed; not
very surprising, too, that inhabitants also witness a pattern of accelerat-
ing local violence at the same time they sense fragmentation on a global
scale. Ironically, this takes place at the same time that the very biological
diversity that moved biologists for much of this century is facing oblit-
eration.

Against such a backdrop filled with mixed emotions, how can we view
the project of unifying biology, and the evolutionary synthesis? Clearly,
the movement toward unity served to exclude and to silence dissenting
voices; and clearly the architects were "white male elites," for there were
few recognizable "others" in their disciplinary network (at that time). Yet
to view the architects (as some would view their contemporaries, the
logical positivists) in the most grim of these politicized labels is to fail to
understand the history (and historicity) of knowledge and the contingencies of their (and our) existence. In upholding their Enlightenment
credo, the "architects" had sought to construct a progressive, liberal, and
secular worldview through their belief in evolutionary humanism; even if
we can no longer share such a project without considerable misgivings,
we would fail miserably as students of history if we could not, or at least
attempt to, understand their project in their own terms. In this historical,
contextual sense there really was a grandeur in their view of life (to use a
phrase much beloved by evolutionists).

No doubt, many readers will inevitably read into this epilogue—and
the text as a whole—whatever sociopolitical perspective their reading
lenses permit them to see; no doubt this text will simultaneously be perceived as positivist and antipositivist, pro-Enlightenment and anti-En-
lightenment, proscience and antiscience, pro-West and anti-West. If it
generates such a range of contradictory interpretations from its readers,
then it will have accomplished one of its goals, albeit a hidden goal thus
far: the stripping of the masked or tacit belief that Western systems of
thought successfully uphold noncontradiction. If there *is* a *logos*, then it is
embedded firmly within its *mythos*; so, too, science is a cultural expression of an aesthetic need, desire, or possibly even hunger or passion,
which attempts to understand the world in order to add some meaning
to life. It is also a story-constituted activity that emerges from and sus-
tains a Western, Enlightenment worldview that through its histories also
demarcates its boundaries. But while an infinity of stories, interpretive

frameworks, and narrative worlds is a possibility, not all stories will do to sustain each and every culture. Science is a culture of its own that meets with its share of evidential constraints, inherited, negotiated, and practiced within a system of undergirding, largely hidden, and most often tacit values. Similarly, the attempt here to reconstruct a narrative felicitous to the perspectives of scientists in *Unifying Biology* is just one of a possible number of stories that meets with its usual share of evidential, narrative, and other such constraints; but because it weaves together such apparently disparate and formerly disengaged narratives it becomes a grand unifying narrative of its own. The possibility of existence without such narratives is difficult to contemplate.

The exegesis of *Unifying Biology* began with the guiding image of William Blake's *Fall of Man*. Returning to this image we can see where the architects of the evolutionary synthesis might have located themselves: toward the center of the text so close to the body of Christ that we may view evolution and religion as descendants of an originary grand, mythic narrative of human origins: small wonder that the two have come in conflict so many times. Anti- or post-Enlightenment, postpositivist, postmodernist critics situated uncomfortably and painfully at the moving periphery of the text have no place for either unifying force or grand theory. Quite rightly so, our century has already been called "The Age of Extremes."[2] Would it be possible to occupy a middle ground between the two and not be torn apart? Or must we invent and imagine "other" worlds of thought devoid of the dilemmas and dichotomies that characterize Western systems of logical thought? According to Susanne K. Langer, questions like these are strong indicators of epoch-altering shifts: "The end of a philosophical epoch comes with the exhaustion of its motive concepts. When all answerable questions that can be formulated in its terms have been exploited, we are left with only those problems that are sometimes called 'metaphysical' in a slurring sense—insoluble problems whose very statement harbors a paradox." Such questions, she continues, "are capable of two or more equally good answers, which defeat each other . . . a choice among them [rival solutions] really rests on temperamental grounds. They are not intellectual discoveries, like good answers to appropriate questions, but *doctrines*."[3] Have we thus formulated the ultimate such questions between mutually exclusive doctrinal ideol-

---

[2] Ibid.

[3] Susanne K. Langer, *Philosophy in a New Key: A Study in the Symbolism of Reason, Rite, and Art*, 3d ed. (Cambridge, Mass.: Harvard University Press, 1957), p. 9.

ogies, so that all answers cancel each other out? If so, can we then see the end of a philosophical epoch? And what will take its place?

A journey that began in the world of contemporary experience thus ends with a return to the "metaphysical origins" of Western knowledge. And there it must remain for the purposes of this text. *Arche* and *telos* are one. A mere scratch to an evolutionist reveals the life-blood of a storyteller, a poet, a philosopher, and maybe even a theologian. Whether we now uphold a narrative of Biblical creation or the more reasoned narrative of evolution, we will find existence within such narrative frameworks that structurate lived experience. Within those narratives, which tell themselves through us, lies the meaning of life.

Select Bibliography

PRIMARY and secondary sources are noted in each specific instance and included within each relevant chapter. The following is a select list of some classic or very recent sources representative of fields in *Unifying Biology* that include comprehensive bibliographies, lists of references, or bibliographic essays, or are otherwise comprehensive in their scope.

HISTORY AND PHILOSOPHY OF EVOLUTION AND BIOLOGY

Abir-Am, P. G. "The Biotheoretical Gathering, Trans-Disciplinary Authority and the Incipient Legitimation of Molecular Biology in the 1930s: New Perspective on the Historical Sociology of Science." *History of Science* 25 (1987): 1–70.

Adams, M., ed. *The Evolution of Theodosius Dobzhansky*. Princeton: Princeton University Press, 1994.

Allen, G. *Life Science in the Twentieth Century*. New York: Wiley, 1975. Rep. ed. Cambridge: Cambridge University Press, 1978.

———. *Thomas Hunt Morgan: The Man and His Science*. Princeton: Princeton University Press, 1978.

Barlow, C., ed. *Evolution Extended: Biological Debates on the Meaning of Life*. Cambridge: MIT Press, 1994.

Bechtel, W., ed. *Integrating Scientific Disciplines*. Dordrecht: Martinus Nijhoff, 1986.

Benson, K. R., J. Maienschein, and R. Rainger, eds. *The Expansion of American Biology*. New Brunswick, N.J.: Rutgers University Press, 1991.

Bowler, P. *Evolution: The History of an Idea*. Rev. ed. Berkeley: University of California Press, 1989.

Browne, J. *Voyaging: Volume I of a Biography of Charles Darwin*. New York: Alfred A. Knopf, 1995.

Burian, R. "Challenges to the Evolutionary Synthesis." *Evolutionary Biology* 23 (1988): 247–69.

Callebaut, W. *Taking the Naturalistic Turn or How Real Philosophy of Science Is Done*. Chicago: University of Chicago Press, 1993.

Cravens, H. *The Triumph of Evolution: The Heredity–Environment Controversy, 1900–1941*. Baltimore: Johns Hopkins University Press, 1988. Orig. pub. in 1978 as *Triumph of Evolution: American Scientists and the Heredity–Environment Controversy, 1900–1941* by the University of Pennsylvania Press.

Cronin, H. *The Ant and the Peacock: Altruism and Sexual Selection from Darwin to Today*. Cambridge: Cambridge University Press, 1991.

Depew, D., and B. Weber, eds. *Evolution at a Crossroads: The New Biology and Philosophy of Science*. Cambridge: MIT Press, 1985.

Gayon, J. "Critics and Criticisms of the Modern Synthesis: The Viewpoint of a Philosopher." *Evolutionary Biology* 24 (1990): 1–49.

Greene, J. *Science, Ideology, and World-View.* Berkeley: University of California Press, 1981.

Grene, M., ed. *Dimensions of Darwinism.* Cambridge: Cambridge University Press, 1983.

Hagen, J. B. *An Entangled Bank: The Origins of Ecosystem Ecology.* New Brunswick, N.J.: Rutgers University Press, 1992.

Harwood, J. *Styles of Scientific Thought: The German Genetics Community, 1900–1933.* Chicago: University of Chicago Press, 1993.

Hull, D. *Science as a Process: An Evolutionary Account of the Social and Conceptual Development of Science.* Chicago: University of Chicago Press, 1988.

Keller, E. F., and E. A. Lloyd. *Keywords in Evolutionary Biology.* Cambridge, Mass.: Harvard University Press, 1992.

Kingsland, S. E. *Modeling Nature: Episodes in the History of Population Ecology.* Chicago: University of Chicago Press, 1985.

Lewontin, R. C., J. A. Moore, W. B. Provine, and B. Wallace, eds. *Dobzhansky's Genetics of Natural Populations I–XLVIII.* New York: Columbia University Press, 1981.

Maienschein, J. "History of Biology." In S. G. Kohlstedt and M. G. Rossiter, eds., *Historical Writing on American Science: Perspectives and Prospects*, pp. 147–62. Baltimore: Johns Hopkins University Press, 1985.

Mayr, E. *Evolution and the Diversity of Life: Selected Essays.* Cambridge, Mass.: Belknap Press of Harvard University Press, 1976.

————. *The Growth of Biological Thought.* Cambridge: Cambridge University Press, 1982.

————. *Toward a New Philosophy of Biology: Observations of an Evolutionist.* Cambridge: Cambridge University Press, 1988.

————. *One Long Argument: Charles Darwin and the Genesis of Modern Evolutionary Thought.* Cambridge: Cambridge University Press, 1991.

Mayr, E., and W. B. Provine, eds. *The Evolutionary Synthesis: Perspectives on the Unification of Biology.* Cambridge: Cambridge University Press, 1980.

Mitman, G. *The State of Nature: Ecology, Community, and American Social Thought, 1900–1950.* Chicago: University of Chicago Press, 1992.

Moore, J. A. *Science as a Way of Knowing: The Foundations of Modern Biology.* Cambridge, Mass.: Harvard University Press, 1993.

Nitecki, M. H. and D. Nitecki, eds. *History and Evolution.* Albany: State University of New York Press, 1992.

Numbers, R. L. *The Creationists: The Evolution of Scientific Creationism.* Berkeley: University of California Press, 1992.

Provine, W. B. *The Origins of Theoretical Population Genetics.* Chicago: University of Chicago Press, 1971.

————. *Sewall Wright and Evolutionary Biology.* Chicago: University of Chicago Press, 1986.

Rainger, R. *An Agenda for Antiquity: Henry Fairfield Osborn and Vertebrate Paleontology at the American Museum of Natural History.* Tuscaloosa: University of Alabama Press, 1991.

Rainger, R., K. Benson, and J. Maienschein, eds. *The American Development of Biology.* Philadelphia: University of Pennsylvania Press, 1988.

Ruse, M. *Darwinism Defended: A Guide to the Evolution Controversies.* Reading, Mass.: Addison-Wesley, 1982.

———. *Philosophy of Biology Today.* Albany: State University of New York Press, 1988.

Somit, A., and S. A. Peterson, eds. *The Dynamics of Evolution: The Punctuated Equilibrium Debate in the Natural and Social Sciences.* Ithaca: Cornell University Press, 1992.

Sterelny, K. "Understanding Life: Recent Work in Philosophy of Biology." *British Journal for the Philosophy of Science* 46 (1995): 155–83.

Waters, K., and A. Van Helden, eds. *Julian Huxley, Biologist and Statesman of Science.* Houston: Rice University Press, 1992.

Webb, G. E. *The Evolution Controversy in America.* Lexington: University Press of Kentucky, 1994.

GENERAL HISTORY AND PHILOSOPHY OF SCIENCE

Appel, T. *The History of the National Science Foundation* (forthcoming).

Ayer, A. J. *Logical Positivism.* Glencoe, Ill.: Free Press, 1959.

Kohlstedt, S. K., and M. G. Rossiter, eds. *Historical Writing on American Science: Perspectives and Prospects.* Baltimore: Johns Hopkins University Press, 1985.

Laudan, R. "Histories of the Sciences and Their Uses: A Review to 1913." *History of Science* 31 (1993): 1–34.

Rheingold, N. *Science, American Style.* New Brunswick, N.J.: Rutgers University Press, 1991.

Thackray, A. *Science After '40.* Chicago: University of Chicago Press, 1992.

Torrance, R. M. *The Spiritual Quest: Transcendence in Myth, Religion, and Science.* Berkeley: University of California Press, 1994.

SCIENCE STUDIES AND CULTURAL STUDIES OF SCIENCE

Ashmore, M. *The Reflexive Thesis: Whrighting Sociology of Scientific Knowledge.* Chicago: University of Chicago Press, 1989.

Bazerman, C. *Shaping Written Knowledge: The Genre and Activity of the Experimental Article in Science.* Madison: University of Wisconsin Press, 1988.

Biagioli, M. *Galileo, Courtier: The Practice of Science in the Culture of Absolutism.* Chicago: University of Chicago Press, 1993.

Buchwald, J., ed. *Scientific Practice: Theories and Stories of Physics.* Chicago: University of Chicago Press, 1995.

Dear, P. "Cultural History of Science: An Overview with Reflections." *Science, Technology, and Human Values* 20 (1995): 150–70.

Fuller, S. *Philosophy, Rhetoric, and the End of Knowledge: The Coming of Science and Technology Studies.* Madison: University of Wisconsin Press, 1993.

Golinski, J. "The Theory of Practice and the Practice of Theory: Sociological Approaches in the History of Science." *Isis* 81 (1990): 492–505.

Graham, L., W. Lepenies, and P. Weingart, eds. *Functions and Uses of Disciplinary Histories*. Dordrecht: D. Reidel, 1983.

Gross, P. R., and N. Levitt. *Higher Superstition: The Academic Left and Its Quarrels with Science*. Baltimore: Johns Hopkins University Press, 1994.

Grossberg, L., C. Nelson, and P. Treichler, eds. *Cultural Studies*. New York: Routledge, 1992.

Hacking, I. *Representing and Intervening*. Cambridge: Cambridge University Press, 1983.

Haraway, D. *Primate Visions: Gender, Race, and Nature in the World of Modern Science*. New York: Routledge, 1989.

———. *Simians, Cyborgs, and Women: The Reinvention of Nature*. London: Free Association Books, 1991.

Holton, G. *Science and Anti-Science*. Cambridge, Mass.: Harvard University Press, 1994.

Landau, M. *Narratives of Human Evolution*. New Haven: Yale University Press, 1991.

Latour, B. *Science in Action: How to Follow Scientists and Engineers through Society*. Cambridge, Mass.: Harvard University Press, 1987.

Latour, B., and S. Woolgar. *Laboratory Life: The [Social] Construction of Scientific Facts*. 2d rev. ed. Princeton: Princeton University Press, 1987.

Lemaine, G., R. Macleod, M. Mulkay, and P. Weingart, eds. *Perspectives on the Emergence of Scientific Disciplines*. The Hague: Mouton, 1976.

Lenoir, T. "Practice, Reason, Context: The Dialogue Between Theory and Experiment." *Science in Context* 2 (1988): 3–22.

Locke, D. *Science as Writing*. New Haven: Yale University Press, 1992.

Myers, G. *Writing Biology: Texts in the Social Construction of Scientific Knowledge*. Madison: University of Wisconsin Press, 1990.

Pickering, A., ed. *Science as Practice and Culture*. Chicago: University of Chicago Press, 1992.

———. *The Mangle of Practice: Time, Agency, and Science*. Chicago: University of Chicago Press, 1995.

Rouse, J. *Knowledge and Power: Toward a Political Philosophy of Science*. Ithaca: Cornell University Press, 1987.

———. *Engaging Science*. Ithaca: Cornell University Press, 1996.

Shapin, S. *A Social History of Truth: Civility and Science in Seventeenth-Century England*. Chicago: University of Chicago Press, 1994.

———. "Discipline and Bounding: The History and Sociology of Science as Seen Through the Externalism–Internalism Debate." *History of Science* 30 (1992): 333–69.

Shapin, S., and S. Schaffer. *Leviathan and the Air-Pump: Hobbes, Boyle, and the Experimental Life*. Princeton: Princeton University Press, 1985.

Stump, D., and P. Galison, eds. *Disunity and Contextualism: Philosophy of Science Studies*. Stanford: Stanford University Press, in press.

Traweek, S. *Beamtimes and Lifetimes: The World of High Energy Physicists*. Cambridge, Mass.: Harvard University Press, 1988.

Woolgar, S. *Science: The Very Idea.* Chichester, Sussex: Ellis Horwood Ltd., 1988.
Wylie, Alison. *No Return to Innocence.* Princeton: Princeton University Press, forthcoming.

HISTORY AND THEORY

Ankersmit, F., and H. Kellner, eds. *A New Philosophy of History.* Chicago: University of Chicago Press, 1995.
Baker, K. M. *Inventing the French Revolution: Essays on the French Political Culture in the Eighteenth Century.* Cambridge: Cambridge University Press, 1990.
Berlin, I. *The Crooked Timber of Humanity: Chapters in the History of Ideas.* Ed. H. Hardy. New York: Vintage, 1992.
Chartier, R. *Cultural History: Between Practices and Representations.* Trans. L. G. Cochrane. Ithaca: Cornell University Press, 1988.
Furet, F. *Interpreting the French Revolution.* Trans. Elborg Forster. Cambridge: Cambridge University Press, 1981. Orig. pub. in 1978 as *Penser la Révolution Française* by Editions Gallimard.
Himmelfarb, G. *The New History and the Old.* Cambridge, Mass.: Belknap Press of Harvard University Press, 1987.
———. *On Looking into the Abyss: Untimely Thoughts on Culture and Society.* New York: Alfred A. Knopf, 1994.
Hunt, L., ed. *The New Cultural History.* Berkeley: University of California Press, 1989.
Hutton, P. "The History of Mentalities: The New Map of Cultural History." *History and Theory* 20 (1981): 237–59.
———. *History as an Art of Memory.* Burlington: University of Vermont Press, 1993.
Kelley, D. R. "What Is Happening to the History of Ideas?" *Journal of the History of Ideas* 51 (1990): 3–25.
La Capra, D., and S. Kaplan, eds. *Modern European Intellectual History: Reappraisals and New Perspectives.* Ithaca: Cornell University Press, 1982.
———. *Rethinking Intellectual History: Texts, Contexts and Language.* Ithaca: Cornell University Press, 1983.
Novick, P. *That Noble Dream.* Cambridge: Cambridge University Press, 1988.
Skinner, Q., ed. *The Return of Grand Theory in the Human Sciences.* Cambridge: Cambridge University Press, 1985.
Veeser, H. A., ed. *The New Historicism.* New York: Routledge, 1989.
White, H. *Tropics of Discourse: Essays in Cultural Criticism.* Baltimore: Johns Hopkins University Press, 1978.
———. *The Content of the Form: Narrative Discourse and Historical Representation.* Baltimore: Johns Hopkins University Press, 1987.

CULTURAL AND LITERARY THEORY

Chatman, S. "What Can We Learn from Contextualist Narratology?" *Poetics Today* 11 (1990): 310–28.

Clifford, J. *The Predicament of Culture: Twentieth-Century Ethnography, Literature, and Art.* Cambridge, Mass.: Harvard University Press, 1988.

Clifford, J., and G. E. Marcus, eds. *Writing Culture: The Poetics and Politics of Ethnography.* Berkeley: University of California Press, 1986.

de Certeau, M. *The Writing of History.* Trans. T. Conley. New York: Columbia University Press, 1988. Orig. pub. in 1975 as *L'écriture de l'histoire* by Editions Gallimard.

Eagleton, T. *Literary Theory: An Introduction.* Minneapolis: University of Minnesota Press, 1983.

Freeman, M. *Rewriting the Self: History, Memory, Narrative.* New York: Routledge, 1993.

Geertz, C. *The Interpretation of Cultures.* New York: Basic, 1973.

———. *Local Knowledge: Further Essays in Interpretive Anthropology.* New York: Basic, 1983.

Langer, S. K. *Philosophy in a New Key: A Study in the Symbolism of Reason, Rite, and Art.* 3d ed. Cambridge, Mass.: Harvard University Press, 1957.

Lentricchia, F., and T. McLaughlin, eds. *Critical Terms for Literary Study.* Chicago: University of Chicago Press, 1990.

Lyotard, J.-F. *The Postmodern Condition: A Report on Knowledge.* Trans. G. Bennington and B. Massumi. Minneapolis: University of Minnesota Press, 1984.

Rosaldo, R. *Culture and Truth: The Remaking of Social Analysis.* Boston: Beacon, 1989.

Said, E. W. *Orientalism.* New York: Vintage, 1979. Orig. pub. in 1978 by Random House, New York.

GENERAL HISTORY

Hobsbawm, E. *The Age of Extremes: A History of the World, 1914–1991.* New York: Random House, 1994.

N.B. Two relevant books became available too late for inclusion, but should also be noted for their comprehensive bibliographic sources:

Depew, D., and B. Weber. *Darwinism Evolving: Systems Dynamics and the Genealogy of Natural Selection.* Cambridge: MIT Press, 1995.

Gayon, J. *Darwin et l'Après Darwin.* Paris: Éditions Kimé, 1992.

Index

Abir-Am, Pnina, 108
Adams, Mark, 26
Adaptationism, 131, 184; critique of, 186, 187; nonadaptive evolution, 130n, 142
"Age of Extremes," 214
Alfred P. Sloan Foundation, post-doctoral fellowships, 195
Allen, Garland, 26, 47, 52, 118; Allen Thesis, 52
American Association for the Advancement of Science (AAAS), 109, 153
American Biology Teacher, 182
American Development of Biology, The, 47
American Institute of Biological Sciences (AIBS), 48, 64, 98, 161, 179, 193
American Journal of Botany, 68–69
American Journal of Plant Biology, 69
American Naturalist, 115, 157
American Philosophical Society (APS), 30, 64, 157
American Society of Naturalists, 48, 65
American Society of Zoologists, 47, 53
anthropological narratives, 87
anthropology, 40; and the synthesis, 198. *See also* postcolonial ethnography
anthropology of knowledge, 90–91
antiphysicalism, 103–4, 205
antisynthesis literature, 34–45
Appel, Toby, 98, 193
aristogenesis, 43. *See also* directed evolution
arrangement of the sciences. *See* ordering of knowledge
Ascent of Man, 164
Association for the Study of Systematics in Relation to General Biology, 140
Australopithecus afarensis ("Lucy"), 40

bacteriology, 163. *See also* cytology; microbiology
Baker, Keith Michael, 90
Barghoorn, Elso, 165n, 196–97
Bateson, William, 145, 147n, 159

"bean-bag" genetics, 174–75
Beatty, John, 54
Beckner, Morton, 105n, 205
behavioral ecology. *See* ecology and evolutionary biology
Bergson, Henri, 111n, 145
Bernal, J. D., 107–8
Biagioli, Mario, 87
Biblical narrative. *See* narrative
biochemistry and biochemists, 52, 66, 194, 196, 206
biological anthropology. *See* anthropology
Biological Sciences Curriculum Study (BSCS), 179–82; textbooks of, 179–80
biological species concept, 30, 134, 136
biology: autonomy of, 100, 104, 106, 109–11, 114, 119, 124, 131, 138, 144, 146, 149–50, 160, 172–73, 176, 186, 205; careers in, 180n; centrality of, 100, 150, 176, 180–81; "crisis" in, 169, 175; departments of, 67; foundations and structure of, 41–42, 105; national contexts of activity, 47; pay scales in, 180n; status, position, location of, 101–2, 181
Biology Council, 69
biometricians, 28
"biosystematists," 153–54
Biotheoretical Gathering (biotheoretical group), 108, 304
Biston betularia, 147
Black, Max, 108
Blake, William, 3–4, 75, 214; *Fall of Man*, 3–4, 75, 214
Bohr, Niels, 151
Bonner, John Tyler, 195
Botanical Society of America, 47n, 68
botany and botanists, 56, 69, 98, 136–37, 154, 199
Brecht, Bertolt, 210
Broad, C. D., 103
Bronowski, Jacob, 164
Browne, Janet, 93
Bumpus, H. C., 120
Butler Act (1925), 66

Langer, Susanne, 12n, 14, 191, 214
Latour, Bruno, 83–84
Leakey, Richard, 40
Lederberg, Joshua, 165n, 172, 197
Lee, Spike, *Do the Right Thing,* 212
Lepidoptera, 120. *See* industrial melanism
Leviathan and the Air Pump, 48, 80–82
Levitt, Norman, 6, 76n; *Higher Superstition: The Academic Left and its Quarrels with Science,* 6, 76n
Lewontin, Richard C., 28, 38, 41, 197
liberalism, 139,145–46, 150, 152, 203. *See also* Huxley, Julian
Life: Outlines of General Biology, end paper, 112–13
Loeb, Jacques, 118
Lyotard, Jean-Francois, 9
Lysenkoism, 51

Mach, Ernst, 14, 103, 118, 203
macroevolution, 34. *See also* microevolution and macroevolution
macromutation, 158
Mainx, Felix, 167–69, 205
marginalization. *See* Goldschmidt
Margulis, Lynn, 196
mathematical population genetics, 27, 28, 30, 57, 68, 119; as core of evolution, 162n; mathematical models, 121. *See also* Fisher, R. A.; Haldane, J. B. S.; theoretical population; Wright, Sewall
Mayr, Ernst, 21, 24–25, 26–30, 52, 61–62, 116n, 118, 134–35, 147, 159, 175–76, 185–88, 192–93, 197, 199, 201–2, 207–9; dispute with Wald, 174; role in SSE and *Evolution,* 57, 64, 153–57; *Systematics and the Origin of Species,* 21, 134–35
mechanism vs. vitalism, 106–7, 109, 144
memory, historical, collective, and disciplinary, 7, 10, 16, 89, 92–93
Mendel Centennial Symposium, 25
Mendel, Gregor, 20, 25–26, 114; Mendelians, 28; Mendelism, 142; rediscovery of, 23, 52
Mendelsohn, Everett, 27, 185
mentality or mentalité, 14. *See also* weltanschauung
Merz, John Theodore, 12
metaphors, selection of, 7–11
metaphysics, in biology and evolution, 100, 129, 132, 160, 169; metaphysical

elements, 100, 103, 111, 114, 123, 130–32, 214; metaphysical stage of development 100, 104, 169, 211
metaphysical origins, 211–15
microbiology, 163, 192, 196–97, 200. *See also* bacteriology; cytology
microevolution and macroevolution, 125, 127n; continuum between, 127–29, 132, 148, 158, 159; Goldschmidt's challenge to, 159; sundering of, 161, 186
Miller, Stanley, 174
model organisms, 56, 126, 147n
modernism, 51; modern period, 99
modern synthesis of evolution. *See* evolutionary synthesis
molecular biology, 36, 50, 52, 66, 70, 174–77, 192, 194, 197, 206
"molecules to man," 178
Monkey Trial, 66
Morgan, Thomas Hunt, 56, 116, 125–27, 140
Morris, Charles, 102, 204
Muller, Hermann J., 23, 157–58, 165, 179
multiculturalism, 5, 212
multicultural theories of knowledge and science, 5, 202

Nagel, Ernest, 152, 176, 205
narrative, 11–12, 16, 87–88, 93–94; Biblical narrative, 3–4, 151n, 215; definition of grand narrative, 9n; grand narrative, 6–10, 99; mythic narrative, 11, 214; narrative "drive," 12–13, 191; narrative pattern, 15; narrative script, 13, 91, 93, 206; unifying narrative, 191
narrativity. *See* narrative
national contexts for evolutionary activity, 33; Anglo-American context, 47, 137, 207–8
National Research Council, 154, 161
National Science Foundation, 69, 179
natural history: decline of , 52, 66–67, 114–19, 125; museums, 55, 67
naturalists, naturalist-systematists, 21, 24–25, 29–30, 66–67, 116, 119, 134–35. *See also* systematists and systematics
Needham, Joseph, 107–8, 204
neo-biology, 163n
Neo-Darwinian synthesis. *See* evolutionary synthesis
Neo-Darwinism. *See* evolutionary synthesis

ABOUT THE AUTHOR

Vassiliki Betty Smocovitis was born in El Mansura, Egypt, of Greek parents. She received her primary and secondary education in Canada, earning an Honors B.Sc. degree in biology with an honors in plant sciences from the University of Western Ontario. In 1988 she received a Ph.D. from Cornell University in the graduate field of "Ecology and Evolutionary Biology," and in the Program for the History and Philosophy of Science and Technology. Her dissertation explored botany and the evolutionary synthesis by focusing on the life and work of G. Ledyard Stebbins, Jr. Following appointment to the department of history at the University of Florida in 1988, she became visiting assistant professor and Mellon fellow in the humanities in the Program for the History of Science at Stanford University in 1990–92. She has also visited the department of philosophy at Emory University as part of a National Endowment for the Humanities Institute, and the section of ecology and systematics at Cornell University. She is presently assistant professor in the history of science in the department of history, University of Florida, where she teaches the history of science in Western culture and the history of modern biological thought.